THE EVOLUTION OF SEX

The Evolution of Sex

Strategies of Males and Females

Kevin Lee Teather

OXFORD
UNIVERSITY PRESS

OXFORD
UNIVERSITY PRESS

Great Clarendon Street, Oxford, OX2 6DP,
United Kingdom

Oxford University Press is a department of the University of Oxford.
It furthers the University's objective of excellence in research, scholarship, and education by publishing worldwide. Oxford is a registered trade mark of Oxford University Press in the UK and in certain other countries

Published in the United States of America by Oxford University Press
198 Madison Avenue, New York, NY 10016, United States of America

British Library Cataloguing in Publication Data
Data available

Library of Congress Control Number: 2024931854

ISBN 9780198886716
ISBN 9780198886730 (pbk.)

DOI: 10.1093/9780191994418.001.0001

Printed and bound by
CPI Group (UK) Ltd, Croydon, CR0 4YY

To my wife Kim, who thinks I'm the most unromantic person she knows.

Preface

As undergraduate students, we were given the option to complete a one-year research project on a specific topic in the fourth year. It was the practice of many professors to let students know what projects they would be interested in supervising, and Dr Donal Hickey, a young faculty member at the time, was interested in the evolution of sexual reproduction. I remember asking myself at the time—Why? As an undergraduate in Biology who had taken courses in Genetics, Animal Behaviour, and Evolution, I wondered what could be left to answer on such a prominent subject. Instead, I decided to pursue a project dealing with spacing patterns (coloniality and territoriality) in mobile animals, a research area that provided necessary background information for my upcoming graduate work. Only later did I realize that the evolution of sex was an important topic with many unanswered questions. Unfortunately, Dr Hickey died recently. I'll never forget him, however, not only for being a great teacher but also as the first person to point out the fact that the evolution of sex still generated substantial interest and was a topic that warranted further investigation.

How sexual reproduction first evolved and eventually became the predominant mode of getting an organism's genes into the next generation remains a much-debated topic in Biology, and is the first objective of this book. It's hard to believe that life on Earth went on for over a billion years without resorting to it, especially given that nearly all multicellular organisms now use it. But this also demonstrates that sexual reproduction is unnecessary for the continuation of life. Reproduction without having to combine genes with another individual seems much more efficient. Why should one go to the trouble of finding and assessing another member of your species to pass on half your genes when asexual reproduction would not require a mate and would result in all your genes being passed on? Sex must carry considerable benefits to overcome all its possible costs.

After discussing the evolution of sexual reproduction, my second objective is to examine the reproductive strategies of males and females. Why males and females arose is, of course, an essential part of sex. But why? Given that there are benefits to sex, why not simply combine your genes with another similar individual? In other words, why should the two sexes be so different? From my research and teaching, as well as life outside of academia, it has become clear that the sexes approach sex in very different ways. In this book, I discuss why there are two very different sexes and why their reproductive strategies are so different. Males and females have an overwhelming drive to contribute their genes to the next generation, and sometimes, they must work co-operatively to do this. But often, the reproductive interests of males and females clash. Indeed, this conflict may even result in one sex reducing the reproductive performance of the other. This approach to sex may seem counterintuitive, but it makes sense in many situations. Thus, in some cases, the goals of males and females align; however, in many cases, they differ.

I have little background in Molecular Biology, and most of my experience in both research and teaching could be classified as 'evolutionary ecology and behaviour'. However, any discussion concerning the evolution of sex and the reproductive strategies of males and females must incorporate current molecular techniques, at least to a certain extent, as they have been instrumental in dealing with specific topics in this field. Any details concerning Molecular Biology are relatively straightforward, and anyone interested in the evolution of sexual strategies, including Biology students, individuals in Evolutionary Psychology, Sociology, Anthropology, and even the general public should find any discussions in these areas readable. On the other hand, experts in these fields may find them oversimplified. Hopefully, I've reached a middle ground.

Although I try to provide a wide range of examples throughout the book, I've not attempted to complete an extensive literature review. Some people have spent lifetimes researching each of the topics covered here and know much more than I do about each. My main purpose was to write a book that covers all the various issues I teach now and wish I knew more about as an undergraduate student. I try to provide enough recent references to illustrate the field's dynamic nature and serve as a starting point for those who wish to pursue particular topics in more detail. As a result, the literature I cite may be slightly biased, not intentionally, but to back up the point or provide an example of what is being discussed.

Unfortunately, despite its importance, the evolution of sexual reproduction and how males and females approach sex are often not covered in detail in most Biology curricula. Such topics may be dealt with superficially in the study of evolution and animal behaviour, but only certain universities provide a specific course designed to address these subjects in detail. One of the problems is that different instructors put more emphasis on different topics, often depending on their own interests and experiences. I've attempted to write this book generally enough that it can be used as a starting point for any courses dealing with the general topic of sexual reproduction in animals.

Many colleagues have commented on specific aspects of the book, including Drs Stevan Springer, Dave Schutler, Kim Critchley, and Christian Lacroix. Christian deserves special thanks for reading all the chapters and providing his thoughts on content and grammar. Not bad for a botanist. I would be remiss if I didn't mention the people who supervised me in various research projects in preparation for my career and played an essential role in my interest in this subject. These include Drs Ralph Morris, Raleigh Robertson, and Pat Weatherhead. When continuing my postdoctoral studies at Cambridge University, I was fortunate to work in the lab of Dr Nick Davies. Finally, with the greatest appreciation, I thank all the students who have attended my classes and generated many interesting discussions about the evolution of sex and the strategies employed by males and females.

Contents

1

Introduction

1.1 Overview

Around four billion years ago, the Earth was hugely different than it is today. The atmosphere was filled with methane, ammonia, hydrogen sulphide, and carbon monoxide, while having very little oxygen. The planet's surface was being bombarded by meteors and exposed to intense ultraviolet light (Figure 1.1). During this time, a molecule formed (or arrived on a meteor) that had the ability to replicate itself. Certain versions could withstand these harsh conditions and were more likely to produce copies that were largely identical to themselves. Still, mistakes were made. Some of the 'offspring' died out as they fared more poorly in those environments. Others were slightly better at surviving and replicating their beneficial characteristics until, finally, the simplest types of single-celled organisms arose. Reproducing oneself has been a property of all organisms throughout time; without it, life forms would not persist for more than a generation. And without that imperfection in the replicating phase, life forms couldn't become better at extracting resources from their environment, escaping other organisms, or even reproducing themselves.

Reproduction is the common thread that links all organisms on Earth. However, reproduction has changed dramatically for most species since those very early days, from one organism simply replicating itself, to two individuals getting together and combining their genetic material to produce offspring. But why did this happen? It seems to be a very inefficient way to reproduce; after all, each resulting progeny only carries half the individual's genetic information, with the other half coming from the individual's partner. And this doesn't include all the additional costs associated with finding a suitable mate with whom to combine their genes. But if we assume that the benefits of two organisms combining their genetic material to produce offspring outweigh the disadvantages, we have another mystery to solve. Why are these individuals so different? We term these two distinct types of creatures 'males' and 'females', and each is trying to replicate themselves as efficiently as possible, even if this is to the detriment of their so-called partner.

So, reproduction today is often quite different than the first form of reproduction. Nevertheless, it is the driving force behind the existence of all living things. Without it,

The Evolution of Sex. Kevin Lee Teather, Oxford University Press. © Kevin Lee Teather (2024).
DOI: 10.1093/9780191994418.003.0001

Figure 1.1 *When life began, the Earth was quite different than today. This drawing is an artist's concept of the young Earth being bombarded by asteroids. Self-replicating molecules evolved into membrane-bound cells, currently regarded as the earliest life forms.*
Image Credit: NASA's Goddard Space Flight Center Conceptual Image Lab.

or even if it occurred at a very low rate, an organism's evolutionary lineage would simply die out. Any organism that combines its genes with another only needs to produce, on average, two surviving offspring over its entire life if the population is to remain stable—one to replace itself and another to replace its partner. This doesn't seem like much, especially when the reproductive potential of some organisms may be in the millions. For example, the giant clam (*Tridacna sp.*) can release more than 500 million eggs at a time and may live for 30 years! Of course, if all the progeny lived, our oceans would be quickly inundated with giant clams, a process recognized by Charles Darwin. But I'm getting ahead of myself.

I have two objectives for this book. The first is to explain why two different organisms of the same species are generally (but not always) necessary for reproduction to take place. This is the problem of the evolution of sexual reproduction; the incentive for the change from self-replication (asexual reproduction) to combining your genes with another (sexual reproduction) has perplexed evolutionary biologists for over 150 years. As I hope to show, sexual reproduction wasn't inevitable, but once it evolved, became ubiquitous among much of life on Earth. The second is to explain why the two parents are usually so different and to understand their often conflicting, but sometimes co-operative, roles in passing on their genes to subsequent generations. In doing this, I rely on several examples from various species although, as I admit later, the examples I use are often, but necessarily, biased.

The terms 'sex' and 'reproduction' are often used together, which frequently leads to confusion about their actual meanings. If we look at them separately, it is clear that these terms refer to two different processes. A further problem is that 'sex' can have a variety of meanings, and it's essential to clarify how it is used. Throughout this book, the word 'sex' can indicate three different things. First, the 'sex' of an organism refers to its biological characterization of being a male or female. It is often associated with things like reproductive organs, genes, and hormones and can be applied to individuals of all sexually reproducing species. However, as I hope to clarify, none of these characteristics define the sexes. There are only two biological sexes—males and females—regardless of the plant or animal we are studying. In some cases, a species may have more than two mating types, but these cannot be considered sexes; I'll touch on this in Section 1.4 and discuss it more fully in Chapter 4. I'm not downplaying the range of gender roles that have been identified in humans, but as I will explain, these are quite different, although often related to the biological sex of a person. Second, 'sex' is also a short form for 'sexual activity', as in 'Last night they had sex', or is often used to infer 'sexual reproduction'. Although 'sexual reproduction', by definition, necessitates a potential increase in numbers, humans having 'sex' may be trying to avoid this at all costs. Sexual activity obviously has nothing to do with the biological sex of the people who engage in this intimate pursuit. In humans and other animals, sexual activity could just as easily occur between two males, two females, or more than two individuals. Finally, the term 'sex' technically infers that there is a combination of genetic material from two cells. Typically, these cells come from different individuals, although this isn't strictly necessary. This last way of looking at sex, when combined with 'reproduction', has been problematic for evolutionary biologists, as described above. In this book, I use the term 'sex' in all three ways and trust that its meaning will be clear from the context.

I should also say that this is not a 'how-to' guide for anyone looking for performance-enhancing tips or solving relationship problems. You won't learn any special skills that will make you a better lover or help you to attract a mate. I won't offer advice on changing your personality, improving your looks, or how to appeal to another using hypnosis (if I knew such techniques, I might have used them myself). In other words, it is not a self-help book. However, gaining a better understanding of sexual reproduction and the strategies of males and females might help readers better address some of the many pitfalls in relationships. Men and women often have different ways of doing things, and it can be frustrating to understand each other. Many of these differences are ultimately related to different evolved strategies to ensure males and females pass their genes on to the next generation. Having said that, I must emphasize that many aspects of sex in humans are strongly overlaid by social elements that affect every part of our lives, and I try to steer clear of these social influences. It's important, though, to determine which behaviours have biological roots and which have been masked by our own social development. So, I don't talk about humans much, although in some instances they make very good examples to highlight a certain point, biologically speaking, that is. And let's face it: we know more about (or at least are more interested in) sex in our own species than in any other. However, when discussing any aspect of human behaviour, it is always important not to fall into the 'naturalistic fallacy', which I explain further in Section 1.5.

Let me point out an inherent bias before going too much further. For a lot of people, the animal kingdom refers to mammals. If they thought about it, many would admit to the odd reptile and amphibian, but (as I've heard frequently) fish are fish; birds are birds; and neither are animals. Also, forget about invertebrates—they are in a different category altogether. Admittedly, how we classify the various living species into broad taxonomic groups is continually changing. The most recent scheme involves placing all living organisms into two superkingdoms—Prokaryota and Eukaryota. These, in turn, are divided into seven separate kingdoms (Figure 1.2).[1] The most recent estimate suggests that the total number of species on Earth, not including prokaryotes for which such estimates are difficult, is about 8.7 million (give or take 1.3 million).[2] Of these, there are about 7.8 million types of animals, while the remaining million or so species belong to the fungi, plants, protozoans, and lesser-known chromists. The best estimate for the number of mammals is around 5500. Mammals belong to the vertebrates, along with reptiles (including birds), amphibians, and fish, for which about 65,000 species have been identified. My point is that mammals, and even vertebrates, make up only a tiny portion of species in the animal kingdom at 0.07% and 0.8%, respectively. The percentages are even lower if we classify them as a percentage of all sexually reproducing species (including nearly all other eukaryotes).

What does the number of species on Earth have to do with the evolution of sex and the strategies employed by males and females? When sexual reproduction is mentioned, along with differences in morphology and behaviour, many people automatically think that humans are a good representative of the animal kingdom (if they indeed think of us as animals). If they are broad-minded, they may think of sex-specific differences in other mammals or birds. However, the animal world is diverse, and there are many different strategies for sexual reproduction used by various species. If we only look at mammals, male elephant seals (*Mirounga sp.*) can be almost four times as large as females, while male and female rabbits (*Oryctolagus cuniculus*) are the same size. Additionally, the males of many mammalian species are ornamented (moose [*Alces alces*] have antlers; lions [*Panthera leo*] have manes; and narwhals [*Monodon monoceros*] have tusks), while many are not (zebras [*Equus sp.*], hamsters [*Phodopus sp.*], beluga whales [*Delphinapterus leucas*]). Of course, there is one characteristic related to reproduction shared by

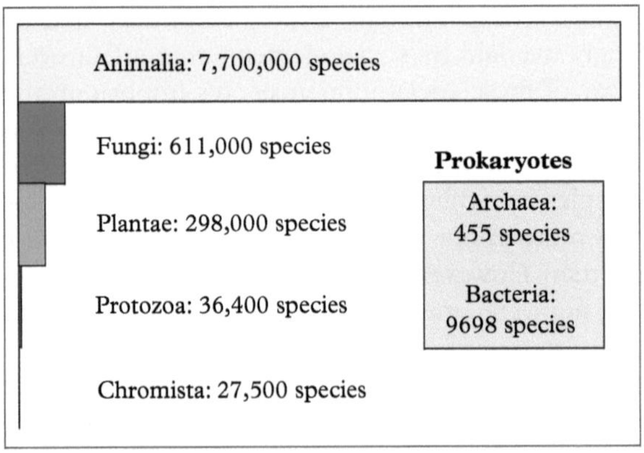

Figure 1.2 *The seven kingdoms of life on Earth and their relative abundance as determined by number of species. Estimates for prokaryotes are a minimum. Species number estimates are from Mora et al.*

all mammals—females feed newborns with secretions from their bodies. However, we would expect, and indeed find, a vast array of differences between males and females in this small group, and this group makes up less than 0.1% of all animals. Since more than 99% of species in the animal kingdom use sex, think of the tremendous diversity of strategies that would be observed if all animals were examined.

There are a few things to keep in mind when reading these chapters. The first I've mentioned: few ideas or hypotheses that are presented use humans as examples to illustrate them. This isn't to say that they don't apply to humans, only that the social aspects of their lives render humans somewhat messy as biological examples. Secondly, I study and teach about vertebrate biology and did most of my research and training looking at mating systems and parental care patterns of birds. Thus, despite all my talk about biases in examining only certain vertebrates as representatives of organisms that reproduce sexually, my examples are heavily weighted toward this group. When possible, however, I try to use examples from other groups, especially invertebrates. In addition, other than a brief overview of their asexual reproduction, I spend little time on plants. This is somewhat unfortunate as plant reproduction is fascinating and quite varied; however, I have mostly left them out due to time and space constraints as well as limitations imposed by my background. In any case, there are many excellent books on reproduction in plants, many from an evolutionary perspective, for example.[3,4] Keep these in mind as you read through the remaining chapters: biases certainly exist, but these biases are probably not that important when discussing the broader issues concerning the evolution of sex and differences between males and females.

After introducing a series of topics in this chapter that assist in understanding the subsequent material, I look at a few ways that individuals reproduce without resorting to sex in Chapter 2. This may appear to be an unexciting topic, especially in today's sex-filled world, but I want to emphasize that many organisms do reproduce asexually. This way of reproducing avoids many problems associated with combining your genetic material with that of another individual, was the only form of reproduction on Earth for a very long time, and has persisted until today. In Chapter 3, I briefly examine the evolution of the eukaryotes from their prokaryotic ancestors and discuss interactions between the DNA of their nuclei and newly incorporated mitochondria. Since the process of meiosis is crucial to sexual reproduction, its evolution is also considered. I then discuss what defines males and females, how they originated in the first place, and the consequences of different gamete size on the reproductive strategies of the two sexes in Chapter 4. In the following chapter, I provide an overview of the varied costs of sex before outlining some of the past and current hypotheses that reveal why sexual reproduction may be so common in Chapter 6. In Chapter 7, I discuss how sex is determined in the first place and situations in which both sexes may occur in one body. An underlying theme throughout this book is how males and females maximize their reproductive success. I more fully explore this in Chapters 8 and 9, examining the two sexes' often conflicting and sometimes co-operative approach to reproduction. But sexual behaviour in eukaryotes isn't just about reproduction, at least directly, and in Chapter 10, I discuss instances when having offspring may not be its sole purpose. Next, I examine how the sexes compete for and select mates in Chapter 11 before addressing the various mating systems that arise from differences in strategies used by males and females to get their genes into subsequent generations in Chapter 12.

The remainder of this chapter is devoted to processes and concepts you should familiarize yourself with before reading further. There is nothing too difficult in the following topics, but understanding them will permit a fuller understanding and prevent many misunderstandings that often arise when discussing sex and the differences between males and females.

1.2 Natural and sexual selection

I deal primarily with evolutionary issues, and, at the very least, a rudimentary understanding of natural selection and evolution is assumed. This book is mainly directed to individuals with such knowledge, and I won't review this process in detail. However, in the event that your background in evolution is weak, here is a brief overview.

Evolution through natural selection was revealed by Charles Darwin, first to his friends, and then, with Alfred Wallace, at a meeting in London in 1858. In short, some individuals have characteristics better than others of their local population in a particular habitat that allow them to survive and reproduce. These characteristics are generally only slightly better than those of conspecifics, so their advantage is usually very small. But even a slight edge is important when competition is fierce. For example, as the climate gets cooler, mammals with slightly more fur would be more likely to survive. And if they were more likely to survive, they would be more likely to have offspring. Given enough time, and assuming that such characteristics have some genetic basis so that they can be passed on from parent to their offspring, the population will gradually change, and organisms will become better adapted to their environments. Of course, nobody expects the environment not to change, at least over long periods, and biological evolution is a continual process. Natural selection resulting in evolution may speed up and go relatively quickly when environments change or remain fairly stable when environments are static for long periods. Other organisms represent a significant part of that environment and are constantly evolving, too. Predators are continually developing better ways to catch prey, and parasites are evolving methods to exploit hosts better. So, evolution through natural selection includes changes in response to abiotic (nonliving) conditions and changes in response to organisms that make up the environment's crucial biotic (living) component.

Evolution through natural selection not only results in organisms becoming increasingly adapted to their environments but also is responsible for the great diversity of living organisms today. This is because natural selection also results in the splitting of lineages so that a single species may become two species. This 'speciation' often occurs when a few individuals find themselves in an environment that differs from that faced by the rest of the population. Both outcomes, an increased adaptation to the local environment and the splitting of lineages, result from the same process—the most suitable organisms are most likely to survive and contribute relatively more of their genes to the following generations.

Does natural selection operate at the level of the gene, the individual, or the group? The role of the individual versus the gene in the workings of natural selection is a much-discussed topic in Evolutionary Biology, and the reader is directed elsewhere if interested in a complete analysis. Instead of dealing with this issue, I'll say that the most successful organisms are those having the highest *fitness*. These are the individuals who optimize the number of high-quality offspring (or genes) in the next generation and maximize their genetic input over the long run. However, I must say something about individual versus group selection for two reasons. First, in the last 25 years of teaching, I have come to realize that a misunderstanding of how natural selection operates often occurs at this level. Second, the advantage of sexual reproduction was, at one point, thought to arise through group selection. Natural selection, except under certain strict conditions, does not operate at the level of the group.

Let me give you an example I often use to illustrate the point to students. Think of an environment that suffers a drought and has insufficient food to feed a population of lemmings (*Dicrostonychini* or *Lemmus sp.*). To minimize this problem, many members of the lemming community sacrifice themselves by jumping over a cliff, freeing up food for the remaining individuals (Figure 1.3). This selfless act would likely give us a sad but satisfying feeling if we were to observe it in a movie. Unfortunately, it's a myth. No lemming in its right mind would sacrifice itself for the group. We only need to consider

Sometimes being the black sheep has its advantages

Figure 1.3 *Lemmings are often used as an example of altruistic behaviour, where they jump off a cliff to preserve a dwindling food supply. But what if one individual cheated? They would be much better positioned to pass on their genes to the next generation. Thus, the requirements for the evolution of altruistic behaviour are restrictive.*

Credit: Martina Zeitler/Copyright: www.justoutsidetheboxcartoon.com.

the genes that would make it into the next generation. Of course, these genes would belong to any lemmings that didn't jump. So, the 'sacrificing genes' are quickly weeded out of the population. In other words, behaviours that benefit the entire group, when at a disadvantage to individuals, are always subject to cheaters. And cheaters, in natural selection, always win.

Even though I expect you understand how natural selection operates, let me clarify a few things. 'Natural selection' is not the same as 'evolution'. Evolution simply means change, and natural selection is one (although the most important) way that (biological) evolution occurs. We know that evolution happens—we can often measure changes directly and often see how things change over time. Natural selection, along with mutation and genetic drift, can change the frequency of alleles in subsequent generations. However, natural selection is the only one of these processes that results in organisms becoming better adapted to their environment. Although there are a lot of misunderstandings about natural selection, usually articulated by individuals who don't have a firm grasp of the subject, it is relatively straightforward and not particularly controversial.

In 1859, Charles Darwin published his classic book describing natural selection, focussing mainly on survival. The individuals who were best able to survive under the environmental conditions to which they were exposed would give rise to descendants who were best adapted to those conditions. It was a simple idea and widely accepted (at least by scientists) at the time. Darwin recognized that males, though, often possessed characters that didn't provide them with any survival advantage. The male moose's (*Alces alces*) antlers, the male cardinal's (*Cardinalis cardinalis*) bright-red colour, and the peacock's (*Pavo cristatus*) ornamental tail are examples (Figure 1.4). Not only could he see no survival value in such characters, but they even seemed to make males more susceptible to mortality. Antlers are energetically expensive to produce, bright-red colouring makes one more noticeable to predators, and bulky tails make flying difficult. So why do they have them?

Darwin suggested that males possess these characteristics because they increase their chances of getting a female to mate. If they do not contribute anything else but sperm to the reproductive process, they need to find a way to mate with as many females as possible. So, males can develop these characteristics for two reasons: either they help them compete with other males for the opportunity to mate with females, or they make them more likely to be selected by females as breeding partners. It's easy to see how certain traits can increase the competitive ability of males. Increased size and weaponry make them more successful during contests with other males. Darwin had no problem with these as he could easily see how such characteristics would increase your chances of breeding with a female. Large males could more successfully compete with other males for access to females. He had more of a problem (as did others) in determining why traits such as the peacock's magnificent tail were valuable (Figure 1.4). In the year following the release of his book describing how natural selection worked, he wrote, '*The sight of a feather* in a *peacock's tail, whenever I gaze at it, makes me sick.*'[5] He assumed they were attractive to females and even instilled in such ornaments an aesthetic value that females could assess.

Figure 1.4 *The male peacock uses his showy tail feathers to convince females to mate. Such characteristics, while increasing the chances of breeding, may actually hinder survival. In this case, the long, showy tail probably makes flying more difficult. Darwin referred to these as sexually selected characters to differentiate them from traits that increased survival. Sexually selected characters are now considered as a type of naturally selected trait.*

Credits: Miroslav Beneda/Dreamstime Photo ID 33648724; DINAL-SAMARASINGHE/Shutterstock Photo ID 2205604017.

Therefore, two types of traits fell outside of Darwin's theory about evolution through natural selection—characteristics found in one sex (usually, but not always, in males) that made them competitive with other members of their sex, and characters found in one sex (again, usually males but sometimes in females) that made them more attractive (for whatever reason) to members of the opposite sex. Darwin concluded that these characters could evolve if they made it more likely to breed for those organisms possessing them. He termed this 'sexual selection' to distinguish such traits from those having survival value, which he placed under the realm of 'natural selection'. According to Darwin, characters that evolved because of sexual selection increased an individual's chances of mating, while those that evolved through natural selection increased their survival ability.

We now recognize that sexual selection is a component of natural selection which includes traits that promote an individual's survival OR reproduction. As usual, Darwin got it right even though he was somewhat perplexed about why one sex should benefit from mating with elaborately adorned members of the other sex. But more about that later.

1.3 Proximate and ultimate questions

When I was a graduate student, I got into a heated argument with a good friend of mine (at least she was a good friend before the argument) who happened to be majoring in Sociology. I can't remember specifically what we were arguing about, but it wasn't until years later that I realized we were debating the same question on very different levels. We were both right but were simply looking at the problem through the lens of our own discipline. No doubt it had something to do with sex, or at least with potential differences between men and women, where amicable discussions can turn into nasty arguments very quickly. It is helpful to know that questions can be answered differently, depending on whether the response is proximate or ultimate. Hopefully, by understanding the difference, you'll be able to avoid some of the pitfalls that are easy to fall into when discussing topics related to sexual reproduction and differences between the sexes.

Proximate questions are often referred to as 'how' questions. How organisms engage in sexual intercourse is pretty evident to most, and there is no reason to go into details here. But suppose we wanted to study it more closely (for purely academic reasons). In that case, we might, for example, look at hormonal changes that occur before, during, and after copulation. Or we might look at heat production in various body regions during sexual arousal and see where differences arise between males and females. Note that these studies don't examine the purpose of having sex. These 'why' questions often fall into the realm of ultimate studies. These look at the evolutionary basis of certain behaviours or structures, or at least why they may be currently maintained, usually by examining their costs and benefits.

A problem often occurs when a 'why' question can be answered proximately or ultimately. If you ask someone why they have sex, you'd probably receive the reply that it feels good. While unquestionably true, this is a proximate explanation. They are really implying that sexual intercourse results in specific chemical changes that activate the pleasure centres of our brains. This answer doesn't tell you anything about its evolution. You could also ask them why they have sex, and they reply that it is the only way to produce offspring. This answer is also correct but is obviously a different way of answering the question.

Let me give you another example that involves sexual reproduction (Figure 1.5). 'Why does the American robin (*Turdus migratorius*) sing in the springtime?' A person giving a proximate answer might provide the following response: 'The days get progressively longer and warmer, and these changes stimulate a surge of testosterone in the males. This testosterone acts on the brain's call centres, which are neurologically connected to the syrinx (the bird equivalent to our larynx). This is why the bird sings in the spring.' 'No', says another person, arguing from an ultimate perspective. 'The robin sings to defend his territory from other males and attract females to mate. The male who is successful at doing both is more likely to breed and leave his offspring in the next generation. That is really why robins sing in the spring.' The first answer attacks the problem from a proximate view, while the second deals with the functional, or evolutionary, roles and is ultimate. But both are right.

I raise this point because, as an evolutionary biologist, I look at questions mainly from an ultimate perspective. This is tricky when you deal with issues that have both proximate and ultimate sides (and can be downright disastrous when arguing with someone who is a sociologist). I'll leave you with one more example. When teaching a class in ecology for non-Science students, very early in my career, a female student accused me of teaching a particular topic (it happened to be mate selection, which I cover in more detail in Chapter 10) in a sexist fashion. I was discussing the various ways males and

Figure 1.5 *Why does the robin sing? Is it because of a surge of testosterone? Or is he trying to attract a mate? Both replies are correct and illustrate the difference between proximate and ultimate answers.* Credit: jbosvert/Adobe Photo ID 236912448.

females assess and select mates. (Oddly, a female friend was accused of the same thing when teaching the same topic at another university.) If you address certain issues from an evolutionary viewpoint, they may seem sexist, but this can largely be blamed on the confusion of proximate and ultimate explanations, as well as to the adherence to the 'naturalistic fallacy' outlined in Section 1.5. The student then showed me her engagement ring and said, 'This is really what girls are looking for.' Her demonstration not only supported my point (that one way, individuals of one sex can assess potential mates is based on resources) but given her criticism, seemed a bit sexist itself!

1.4 Sex versus gender

Since this book deals with the evolution of sexual reproduction, it is understandably based firmly in Biology. As noted earlier, I attempt to stay clear of social issues for which I have little background and, in fact, spend most of the book discussing reproduction in organisms other than humans. But humans are animals and share an evolutionary past with other animals; therefore, I use them as examples when warranted. Additionally, rightly or wrongly, the topic of sex often comes back to humans. I will later make the case that there are only two biological sexes and discuss how they differ from a biological viewpoint. However, one could argue that this omits people who don't fit neatly into this binary division of male/female categories. There are two issues here, both related to determining sexual identity in humans, and I want to deal briefly with both to avoid any misconceptions. I also come back to this topic in Chapter 4 because, above all other topics, it can lead to unnecessary arguments, criticism, and claims of gender bias.

A student approached me at the beginning of a semester recently and told me they go by the name of 'Julian' even though they had been recorded as 'Julie' in university documents. Their sexual identity was different than their biological sex. Other students may perceive themselves as neither male nor female and argue that these are just categories into which we slot people. How individuals perceive their own sexuality is relatively fluid in humans, making defining males and females seemingly problematic. However, this problem exists mainly because of the confusion between sex and gender; thus, it's essential to understand the difference.

As I mentioned before, the behaviour of humans must be interpreted against both a biological and social background. 'Gender' refers to the social categories of individuals and is determined mainly by psychological processes and role expectations. For example, women in Canada make significantly less money than men do for equal work. This problem is a gender issue, not a sex issue. An individual's 'gender identity' is what gender they consider themselves to be, if any. The gender of an individual is generally similar to their biological sex; if an individual has ovaries and a uterus, they most often think of themselves as female. However, in many cases, their sexual identity doesn't align with their biological sex. A biological male may consider himself masculine, feminine, or neither, even though he has fully functional male reproductive organs. So, 'males' and 'females' are used in Science to denote biological sexes and can be clearly separated based on one particular characteristic (described in Chapter 4). On the other hand, 'masculinity' and 'femininity' are social constructs, used often (at least in humans) to characterize the genders of men and women.

1.5 Naturalistic fallacy

I've mentioned the naturalistic fallacy a couple of times and how the lack of a good understanding of it can lead to many problems when discussing the potential biological roots of a behaviour. First described by the well-known philosopher David Hume, and later defined more clearly by G. E. Moore in the *Principia Ethica*, the naturalistic fallacy infers that moral justification for a thought or action does not logically follow simply because it arose naturally. In other words, what is natural isn't necessarily right (or what is right isn't necessarily what is natural). We often see behaviours in other animals that we might think of as distressing because we tend to apply our human concept of morality to these actions. If a cat brings a dead (or partially alive) bird to the door, it often upsets us because our human morality and the cat's actions don't align.

Let me give you another example. Lions live in prides consisting of related females, any offspring dependent on them, and one or a few unrelated (to the females) adult males (Figure 1.6). The related members may be sisters, mothers, grandmothers, aunts, nieces, and daughters, and young dependent males, all numbering a dozen or so individuals. One adult male from another pride, often one member of a coalition of two or three brothers or cousins, usually has sex with all the reproductively active females and is the father of any offspring produced. When the male offspring are old enough and can fend for themselves, they leave the pride and live a nomadic lifestyle for a few more years, wandering with their similarly aged male relatives, getting larger and stronger. When they feel the time is right, they try to take over another pride, often fighting the older males for access to all the females. What happens next is the point of the story. If the younger males are successful at usurping the pride, they often kill all the male and female offspring that are dependent on their mothers. From an evolutionary viewpoint, this is an understandable reproductive strategy—why protect and raise a bunch of offspring that aren't genetically yours? The best way to maximize your fitness is to reproduce yourself and invest in your own offspring. After the offspring are killed, the females come back into breeding condition and can be impregnated by one of the new males.

Figure 1.6 *Males that take over a pride of lions often kill the dependent offspring. Doing this prevents the new breeding male from investing in progeny that isn't genetically his and brings the adult females back into oestrus so he can fertilize them.*

Credit: Santiago Rodriguez/Dreamstime Photo ID 53708634.

Although killing the offspring may seem like an awful thing to do and would be difficult to watch (a video of an adult male lion killing a cute little cub can be quite disturbing), I think that most people would understand it and realize how this behaviour could evolve by natural selection. After all, they are just lions. Just as our cat bringing a bird to the door follows its own instincts, a male lion is just doing what comes naturally to him. However, in humans, infanticide is also much more likely to be carried out by unrelated (step) fathers than by biological fathers.[6] Although possibly influenced biologically (not ruling out any sociological explanations), infanticide in humans cannot be accepted as the morally correct thing to do, regardless of the reason behind it. Too often, people deny the biological roots of a behaviour because they are worried that an attempt to explain a behaviour biologically implies that it is morally acceptable. But the two are not connected. And uncovering the root causes of a morally unacceptable behaviour may help us better address it.

We come back to differentiating between Sociology and Biology. Humans, as a society, must determine right and wrong. This may differ between different cultural groups or different religions. Given different sociological inputs, moral values can change both geographically and temporally. Biological characteristics are similar in all cultures. Sexual reproduction is a principal component of evolutionary fitness, so behaviours related to it might be expected, at least in many cases, to be rooted in biology. However, this doesn't mean that everything related to sex is morally justified—this would be a view rooted in the naturalistic fallacy.

1.6 Summary

The evolution of sex and the resulting strategies used by males and females to maximize their contribution of genes to the following generations has been a fruitful source of arguments, hypotheses, debates, and studies since Darwin. The main objective in writing this book is to provide a general overview of these topics that can be read easily by Biology students and people in other disciplines. Even those outside academia with a general interest in sexual reproduction may enjoy the content as provided from an evolutionary point of view. Although I assume readers have at least a rudimentary understanding of certain topics, I use this first chapter to examine how the term 'sex' is used and briefly review evolution, natural selection, proximate and ultimate questions, the meaning of gender, and the naturalistic fallacy. Some of these may seem unrelated to sex, either its evolution or the strategies used by males and females. However, a general background for each of these topics is necessary to understand subsequent material better and avoid potential misunderstandings later. Although most of the book is not very controversial, disagreements over specific issues are to be expected. Such disputes are welcomed when they stem from biological or evolutionary analyses, such as a different interpretation of the literature. However, readers that base their objections on the social values of humans, or a misunderstanding of the type of question being addressed, are encouraged to re-examine the topic through an evolutionary lens.

References

1. Ruggiero MA, Gordon DP, Orrell TM, Bailly N, Bourgoin T, Brusca RC, et al. A higher level classification of all living organisms. PLoS ONE [Internet]. 2015 Apr [cited 2022 Aug 10];10(4):e0119248. Available from: https://doi.org/10.1371/journal.pone.0119248
2. Mora C, Tittensor DP, Adl S, Simpson AG, Worm B. How many species are there on Earth and in the ocean? PLoS Biol [Internet]. 2011 Aug [cited 2022 Aug 10];9(8):e1001127. Available from: https://doi.org/10.1371/journal.pbio.1001127
3. Barrett SC (editor). Major transitions in flowering plant reproduction. Chicago: University of Chicago Press; 2008. p. 216.
4. Cohen S (editor). Introduction to plant reproduction. New York: Syrawood Publishing House; 2018. p. 256.
5. Darwin C. Letter to Asa Gray, dated 3 April 1860, In: Darwin F, editor. The life and letters of Charles Darwin. New York and London: D. Appleton and Company; 1911. Vol. 2, pp. 90–91.
6. Daley M, Wilson M. The truth about Cinderella: A Darwinian view of parental love. New Haven, CT, and London: Yale University Press; 1999. p. 80.

2

Not All Reproduction Involves Sex

2.1 What is asexual reproduction?

As a young boy, I was often told that cutting an earthworm in two would produce two fully formed earthworms. I probably even did this, although I wouldn't encourage it now. In any case, it won't. Earthworms don't reproduce asexually; they have both male and female reproductive organs, and although they could fertilize themselves, they usually don't. Rather, earthworms reproduce sexually; they must combine their genes (located in the DNA of their cells) with another individual to increase their numbers. Although there may be an increase in parts, there is no increase in the number of individuals after cutting them in half; thus, it is not a form of reproduction. On the other hand, if we do the same thing to a planarian, a type of flatworm often used in Biology classes, both body parts will develop into new individuals (Figure 2.1). In fact, if we were to cut the planaria into several pieces, all of the fragments have the potential to regenerate their missing parts and form complete individuals. In this case, since two or more individuals arose from the parent without any other organism participating, it is asexual reproduction (although unintentional). To be clear, asexual reproduction involves an increase in the number of organisms arising from a single individual, while sexual reproduction refers to increasing the number of individuals by combining the genetic information from two cells. These cells can be from the same organism but are most often from two different parents.

I'll more fully discuss the advantages and disadvantages of both asexual and sexual reproduction soon, but expect that you can see the main problem from an evolutionary standpoint. Generating offspring without involving any other individuals can be a real advantage. If the success of natural selection is based on the contribution of genes to the following generations, why would an organism combine its DNA with that of another individual during reproduction? It would be much more efficient to produce offspring having 100% of its genes. Plus, think of the time saved finding and sizing up potential mating partners to ensure that they were worthy of contributing half their genes to the offspring.

The earliest organisms on Earth were prokaryotes—bacteria and archaeans—that only reproduced asexually. The ability to reproduce sexually was later present in the early eukaryotes (organisms that have membrane-bound cell organelles and include all

The Evolution of Sex. Kevin Lee Teather, Oxford University Press. © Kevin Lee Teather (2024).
DOI: 10.1093/9780191994418.003.0002

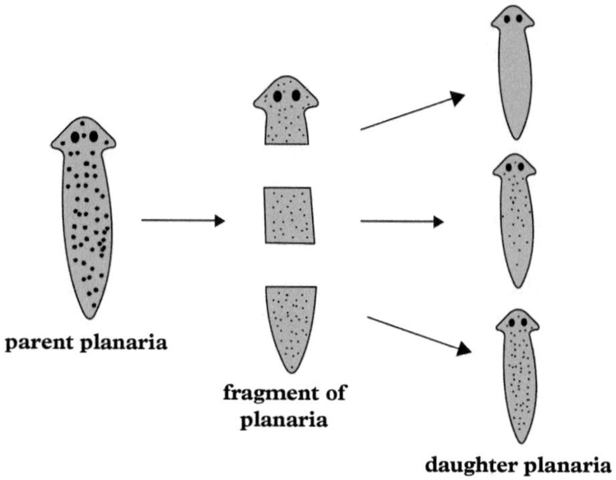

Figure 2.1 *Planaria can regenerate any body parts that are lost. For this reason, if you cut a planarian into pieces, each segment has the ability to regrow the missing parts and become an entire individual. In the case shown in this figure, one individual becomes three individuals after splitting; thus, it is a form of reproduction.*
Credit: Eli-E-learning/Shutterstock Photo ID 1490183600.

protozoans, plants, animals, and fungi) and today is observed in more than 99% of eukaryotic organisms. This places the age of sexual reproduction at a minimum of two billion years. However, life evolved at least 3.8 billion years ago. In other words, asexual reproduction was the only way to contribute to the following generation's gene pool for a long time, almost half the time life has existed on Earth. That's not to say that prokaryotic organisms didn't have sex—they did (and do). Many bacteria today exchange genetic information without an increase in numbers and almost certainly did so back then. That's sex without reproduction. However, asexual reproduction was the original way of contributing an organism's genes to the next generation, was around for a long time, was extraordinarily successful, and still occurs in all prokaryotes, archaeans, and a limited number of eukaryotic organisms today. So why did sexual reproduction evolve in the first place? And why has it been so successful?

Before going any further, I should point out that many eukaryotic organisms that reproduce asexually also do so sexually; the different breeding modes most often happen under different environmental conditions, usually on a seasonal basis. As we will see, these species have been instrumental in better understanding the pros and cons of both methods of reproduction. To provide a possible clue as to what these organisms reveal, consider this. Asexual reproduction may be a good short-term strategy as it produces many offspring rapidly. One avoids the hassle of searching for a mate and instead simply makes clones of itself. This is great when conditions are stable, and you want to have as many offspring as possible in a short period of time. However, it may not make a suitable long-term strategy as the reduced genetic variation in these lineages does not provide the options necessary for a changing environment. If conditions change (which they always do), organisms are less able to adapt if they can only reproduce asexually. But this is only part of the picture. Although nearly all eukaryotic organisms reproduce sexually, some have actually returned to asexuality in the form of parthenogenesis, a reproductive mode that is dealt with more fully later in this chapter. In other words, for some reason, asexual reproduction has been selected over sexual reproduction as a strategy in these organisms.

Let's face it—asexual reproduction seems somewhat uninteresting, at least when thinking about it from a human perspective. After all, the fun and struggles of reproduction often involve the relationship one has with another. And because so much of our lives is dictated by events leading up to, during, and after sexual reproduction, asexuality might also seem somewhat alien to us. So, let's examine the various methods used to reproduce asexually before reviewing the costs and benefits of asexual and sexual reproduction.

2.2 Kinds of asexual reproduction

2.2.1 Fission

Fission, in which the parent splits into equal parts to produce two or more daughter cells, is the oldest and simplest type of asexual reproduction and is found in most bacteria, archaeans, and many single-celled eukaryotes. Binary fission, which results in the production of two daughter cells, is often classified on the orientation of cell splitting—longitudinal, transverse, oblique, or irregular—although this is unlikely to be of interest to anyone other than the organism itself.

Let's look at the typical sequence of events in a bacterium (Figure 2.2). Bacteria are simple creatures, and their genetic material (DNA) is not contained within a nucleus like eukaryotic cells but rather floats freely in a chromosomal mass in the cytoplasm. This mass must be unwound and takes on the shape of a circular structure (roughly) before it is duplicated. At this point, the cell begins to divide, with each daughter cell containing one of the copies of DNA. Presto—two cells from one, without any exchange of genetic information. Such a process can also occur in eukaryotes but is a little more complicated due to the membrane-enclosed structures in the cell. However, it is common in *Euglena*, for example, a well-known single-celled eukaryote that we often see in a drop of pond water magnified under a microscope.

This process is taken a step further in multiple fission, in which the parent divides into many parts of similar size by repeated mitosis. The newly produced daughter cells are often encapsulated in cysts and then released. Amoebae, known to most people, biologists or not, can reproduce asexually by binary fission and, when conditions are unfavourable (e.g. as they travel through the colon of their host), by multiple fission. In such cases, the amoeba first becomes very round and encapsulates itself, forming a cyst. It then undergoes mitosis several times, creating many daughter cells that reside within the cyst. When food becomes more abundant or favourable environmental conditions return, the cyst breaks open, releasing all the daughter cells. Perhaps the most

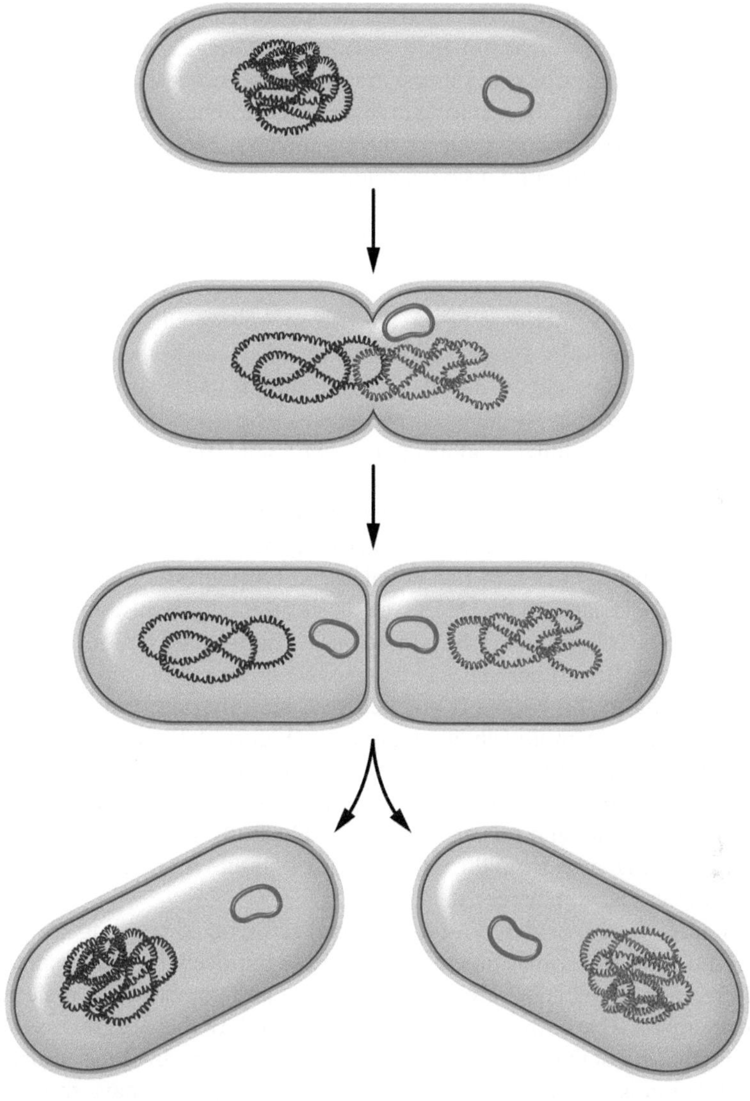

Figure 2.2 *Binary fission in bacteria. After the duplication of genetic material, a septum divides the cell in two, and the newly formed daughter cells ultimately split. Although comparable to mitosis, binary fission has several differences, the main one being the division of the cell while the genetic material replicates.*

Credit: Aldona Griŝkeviĉienė/Dreamstime Photo ID 241462,683.

harmful organism that can reproduce using multiple fission is *Plasmodium*, the single-celled eukaryote that causes malaria. Malaria is among the most common diseases in many regions, accounting for 241 million cases in 84 countries in 2021.[1] An estimated

619,000 of these died. After being released as sporozoites (a specific life stage) into humans (and other types of hosts) by a mosquito, *Plasmodium* reproduces asexually through multiple fission, producing many merozoites (another life stage). These merozoites infect the red blood cells, resulting in fever, muscle aches, headaches, and a variety of other symptoms exhibited by patients having malaria. Merozoites are then ingested by mosquitoes and pass through a sexual stage.

2.2.2 Budding

A new organism can sometimes be created from only a small part of the parent. Often, this region has to be very specific, but in other cases, the new individual can form from just about any part of the parent. Budding is characteristic of a few single-celled organisms belonging to bacteria, archaeans, yeast, and protozoans. It can also be found in many multicellular animals, particularly cnidarians (anemones, corals, and jellyfish).

Yeasts are often classified based on their method of asexual reproduction—budding or fission. The main difference is the initial size of the daughter. When fission occurs, the original yeast cell divides into two equal parts. During budding, a small region of the parent becomes isolated. The DNA was duplicated before this, and one of the two daughter nuclei travels into the newly formed bud. This isolated region, or growing bud, then separates from the mother cell. Hydras, sessile creatures related to the jellyfish we often see swimming in the water off our beaches, are often used in Biology classes to demonstrate budding in multicellular organisms (Figure 2.3). The hydra reproduces quickly in favourable conditions, producing new buds every few days. These buds generally originate in the middle of the body. However, the central body cavity extends into the bud, and the tentacles and mouth soon form. The buds mature, grow into miniature individuals, and break away from the parent. One often sees more than one bud growing off the same individual concurrently.

Strobilation is often considered a more specialized type of budding (or fission) that occurs in some animals, such as tapeworms. In these multicellular organisms, the duplicated body segments (not cells) are called *proglottids* and contain both male and female reproductive organs. The body often consists of hundreds of these segments that mature, separate from the end of the individual, and form brand-new individuals; thus, reproduction by strobilation is asexual. However, because the proglottids have the reproductive organs of both sexes, they can also reproduce sexually.

2.2.3 Fragmentation

The life of any organism is challenging and, as you might imagine, losing part of one's body can make things even more difficult. So, the ability to be able to regenerate missing body parts would be a great help. For example, let's consider an organism that is attacked by a predator. Of course, if its tail happens to be in the stomach of a predator, the regeneration of the head to go along with the rest of the body wouldn't do much

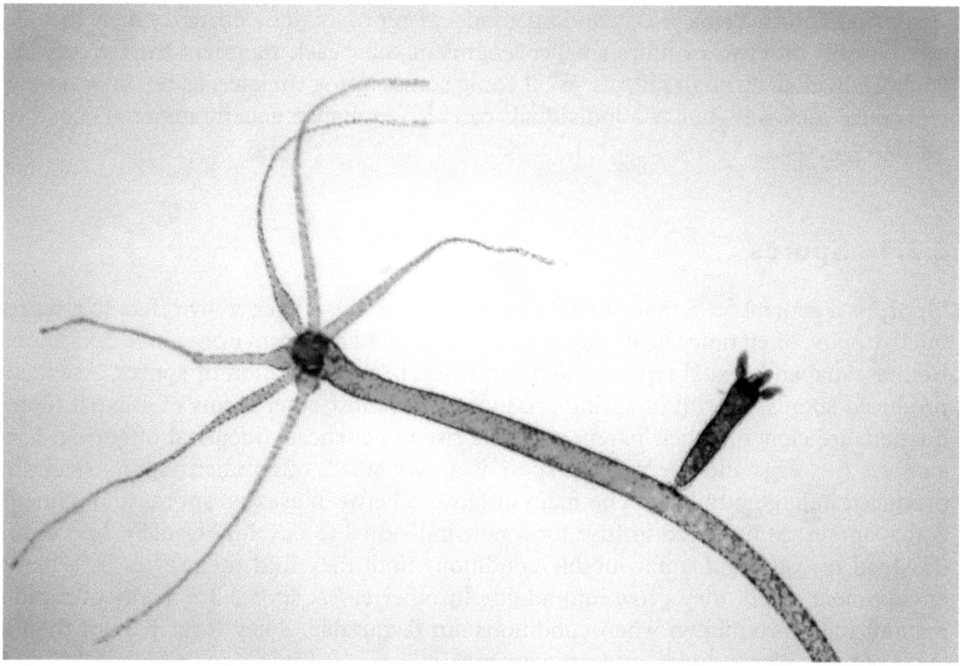

Figure 2.3 *Budding in Hydra. The small bud emerges from the stalk of the adult because of repeated mitotic divisions. When mature, the bud separates from the parent individual and becomes an independent organism. This new organism is genetically identical to its parent.*
Credit: tonaquatic/Adobe Stock Photo 281539046.

good. However, the regeneration of the tail by the hopefully still free-living but tailless individual would undoubtedly be beneficial. Of course, this isn't reproduction, as we would still only have one individual. But what if the organism is cut into two parts in an accident, and both parts grow the missing regions? Two individuals are created in this case, and the population size increases.

Remember the planaria that I talked about at the beginning of this chapter? In this organism, the parent can be split into various parts, and each of those parts will grow into a complete individual. Although this process has received a great deal of attention by the medical community (wouldn't it be great to have the ability to regenerate lost body parts?), the extent to which reproduction occurs through fragmentation in the wild isn't clear. Possibly, reproduction by fragmentation may be more unintentional in this group, with separated body parts often resulting from injury or predation.

Voluntary fragmentation is, however, an essential method of reproducing in many colonial or multicellular groups, including algae, fungi, plants, sponges, flatworms, sea stars, and some annelids. The green, filamentous green algae known as *Spirogyra* are multicellular organisms whose spirally wound chloroplasts give the species its name.

Fragmentation is *Spirogyra's* standard mode of reproduction, during which the filament breaks into two or more smaller lengths of cells; each fragment then grows into a new individual through mitosis. We'll come across *Spirogyra* later, as not only do they reproduce asexually, but two individuals can also exchange genetic material (i.e. have sex).

2.2.4 Spores

'Spore' is a general term that can have multiple meanings. It generally refers to a reproductive body, often unicellular, that is able to withstand harsh environmental conditions. Both asexual and sexual reproduction can result in the formation of spores. Asexually produced spores, as with offspring produced from most other forms of asexual reproduction, are clones of their parent and give rise to genetically identical offspring. Like gametes (i.e. eggs and sperm), the spore is a very small, often microscopic, structure produced in large numbers. The main difference between asexual spores and gametes is that spores do not need to fuse for a new individual to develop. Usually, spores are designed to withstand unfavourable conditions until they find themselves in a better environment where they grow into adults. In other cases, spores are a way of rapidly maximizing reproduction when conditions are favourable. They form part of the life cycle of some plants, fungi, and protozoans.

Algae often produce asexual spores when conditions are favourable, allowing them to increase in numbers quickly. When conditions are unfavourable, however, the algae reproduce sexually; the male and female gamete fuse into a zygospore that withstands harsh conditions. In many cases, spores develop into sporangia, structures that are well-preserved in fossils in one of the oldest eukaryotes, a 1.2 billion-year-old red alga.[2] Mosses also have a reproductive cycle that involves both sexual and asexual stages. The sporophyte typically forms spores by halving the number of chromosomes through meiosis. These are released into the air, usually in dry or harsh conditions. A small proportion of these will land in suitable areas and begin the photosynthetic gametophytes development phase as either male or female. This is the stage that we generally think of as moss. The male gametophyte eventually releases sperm, which makes its way to the female gametophyte. There, fertilization occurs; the resulting sporophyte begins to grow; and the cycle starts again. If you turn over a fern leaf, you may see several small dark spots; these are the sporangia that produce spores. The spores are released by the fern and dispersed by the wind; if they land in a suitable location, the spores grow into gametophytes. In this case, the gametophytes look nothing like the ferns we are used to seeing. Instead, they produce male and female reproductive organs that generate male and female gametes. Once these fuse, the new plant develops into the plant we know of as a fern.

When we think of spores, we often think of illnesses, especially diseases of the respiratory system caused by breathing in large numbers of them. Most often, these are the spores of fungi. Nearly all fungi produce spores sexually or asexually. In fact, the concentration of fungal spores in the air is generally 100 to 1000 times that of pollen,[3] contributing to 1500 deaths per year associated with asthma in the UK.[4] Fungal spores come in many shapes and sizes and become airborne when agitated by slight movements of air or water currents (Figure 2.4). Those of the more advanced fungi require that spores (known as meiospores) of different mating types come together; reproduction, in this case, is sexual. More commonly, asexual spores (called mitospores) are produced by one parent through mitosis, resulting in an individual identical to the parent.

Figure 2.4 *The most common way of reproducing in fungi is through asexual spores. The spores released by this puffball are formed through mitosis and can be used to colonize new habitats. Certain fungi also reproduce asexually using budding or fragmentation. Those species that reproduce sexually often do so in response to adverse environmental conditions.*

Credit: Sergey Kichigin/Dreamstime Photo ID 62234450.

2.2.5 Vegetative

Plants are masters of asexual reproduction. Indeed, plants are so variable in how they reproduce, that one could easily fill a book just discussing them. Vegetative reproduction incorporates many of these ways, where a group of cells form a new individual genetically identical to the parent plant. Many plants, such as poplar trees and dandelions, send out underground suckers, or rhizomes, and a clonal population may soon cover the area. Alternatively, some plants send stems above the ground. For example, strawberry plants employ runners or stolons, which are stems that emerge from leaf nodes and form roots in suitable areas for new plants. The aquatic duckweed develops plantlets on the margins of its leaves, a form of reproduction seen in many other plants, such as cacti. Bulbs (e.g. onions), tubers (e.g. potatoes), and corms (e.g. crocuses) all employ the method of dividing underground structures into more individuals. Finally, we can help new individuals develop in our gardens or farms by using cuttings or grafting individuals together. While a detailed discussion about how plants reproduce asexually falls outside the scope of this book, it's important to realize that many species in this group can multiply without resorting to sex.

2.3 Asexual reproduction by cells in an organism

Of course, asexual reproduction happens all the time by cells within an organism. Indeed, all of the cells in the body of a multicellular organism are descended from a common ancestor (or two if we include the genetic material from the male contributor of a sexual union). Cells differentiate (e.g. muscle, bone, and nerve) by activating different genes even though they all have the same genetic coding. However, individual cells of any type are subject to minor changes resulting from mutation or DNA replication errors. This process leads to a certain amount of genetic variation within the cells of a body and is, therefore, subject to natural selection. In a way, the body's cells are analogous to lineages on an evolutionary bush.

I recognize there are obvious differences between the reproduction of individuals and the reproduction of cells in their bodies, and I raise this issue only because of its importance in cancer studies. Not to get too far off-topic, 'cancer' is defined as the uncontrolled reproduction of cells. Increased mutation rates are characteristic of most cancer cells, and there is much that evolutionary studies can tell us about tumours and treatment. Suppose the original cancer has spread and resulted in different tumours in the body. In that case, we can usually track the lineage of tumours by looking at their genetic variability. Higher genetic diversity is expected in the original tumour because DNA in the cells has had a longer time to mutate.[5] Conversely, the most recent tumours should have lower genetic diversity because the DNA has had a shorter time to mutate. Of course, this depends on the number of cells that have 'colonized' subsequent tumours. The process is similar to what we'd expect in the genetic diversity of populations spreading between islands.[6] It is just one of the reasons why all medical doctors should have a fundamental understanding of evolutionary biology!

Finally, although one often hears that asexual reproduction has not occurred in humans (let's avoid discussing how the Virgin Mary reproduced), it does happen with some regularity. An egg, once fertilized (of course, the whole process must begin with sexual reproduction), can split and give rise to two individuals who develop into monozygotic twins. Thus, we have doubled the population (from one to two) without any (further) input from another individual. The resulting two individuals are genetic clones of each other but not identical to their mother.

2.4 Sex without reproduction: Gene transfer

If your offspring were always the same, natural selection wouldn't have much with which to work. Reproducing copies of yourself might be good if the parent was successful and environmental conditions were identical for the offspring. In other words, if it works, don't mess with it. However, change is often good. Successful individuals might be even more successful if they had certain characteristics already present in others. And the environment may differ slightly for offspring because conditions are likely to change temporally or geographically. So, there must be a way, besides mutation in the DNA (which happens very slowly), for incorporating new genetic material into one's body, even if reproduction is asexual.

Horizontal or lateral gene transfer was, and remains, an important process that allows genes from one individual to move into another. It happens between two members of the same or different species and can be responsible for rapid evolutionary changes. It was initially believed that horizontal gene transfer only occurred among various prokaryotes, but it has since been discovered that genes can be transferred between a range of species, even bacteria and humans! However, let's look at how genes are shared between closely related prokaryotes since they have no other way of quickly making significant changes to their genome.

There are three ways in which genetic material can be transferred between prokaryotes (Figure 2.5). 'Transformation' is the uptake and incorporation of free-floating DNA into a recipient bacterium. There are several steps required for transformation to occur—the release of DNA from the donor cell (generally because of death and degradation), the persistence of DNA outside of the cells, the development of competence by the host cell, the uptake of DNA, and the integration of DNA into the host cell genome. The development of competence, or receptivity, suggests that the host cell must be physiologically primed before taking up the genetic material. Transformation generally occurs with a change in an environment associated with increased population density or low food conditions. In other words, it takes place when it might be a good idea to change. The process of transformation, as will be discussed more fully in the next chapter, may have been instrumental in the development of sexual reproduction.

'Transduction' is a second method to transfer genes laterally, and it involves a bacteriophage (a virus of bacteria) transporting the DNA from one bacterium to another.

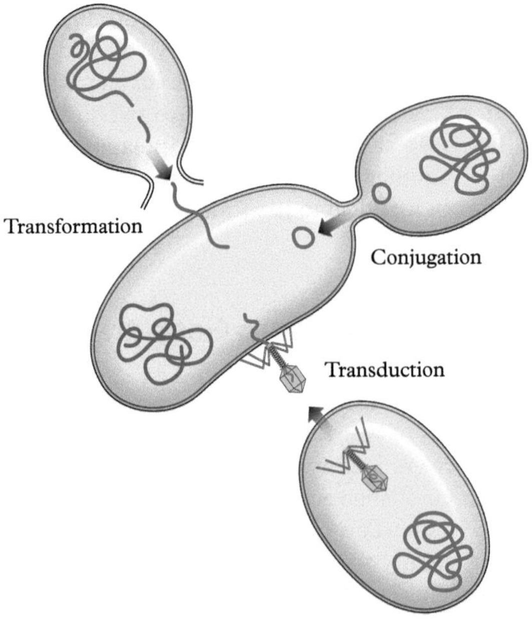

Figure 2.5 *Bacteria can use three ways to increase their genetic variability. First, they pick up a strand of DNA released from dead bacteria and incorporate it into their own genetic material (transformation). Second, they make contact with another bacterium and obtain a segment of DNA through a conjugation tube (conjugation). Third, a bacteriophage (a virus that infects bacteria) is used to carry DNA from one bacterium to another (transduction).*

Credit: Aldona Griŝkeviĉienė / Dreamstime Photo ID 241462681.

In this case, the bacteriophage first infects the host bacterium, hijacking its metabolic machinery to produce more phages. When the host cell dies, one or more bacteriophages may carry a part of the host bacterium's DNA. This phage then invades another bacterium, releasing the DNA which may be incorporated into the host bacterium's genome.

Thirdly, 'conjugation' involves the transfer of DNA from one bacterium to another involving direct contact or a conjugation tube (or pilus) between cells. Conjugative plasmids, small segments of DNA in the donor bacterium known as F plasmids (for fertility), carry the genes, promoting the pairing process in addition to the genes that are transferred from one bacterium to another. Briefly, a conjugative plasmid in the donor cells stimulates the formation of the tube between cells, pulling both cells closer together. A single strand of the double-stranded plasmid DNA is transferred into the recipient cell through the pilus, so both the donor and recipient cells have a single copy of the DNA. The plasmid DNA replicates in both cells, so they each have complete copies, and both can now form conjugation tubes with other bacteria. Conjugation can occur between bacteria of the same or different species.

2.5 Is parthenogenesis a type of asexual reproduction?

The New Mexico whiptail lizards (*Aspidoscelis neomexicanus*) (Figure 2.6) are commonly found scurrying about the drier areas of the Southwestern United States, where they feed primarily on insects. A medium-sized lizard, they have achieved a certain prominence as the state reptile of New Mexico. During the breeding season, one lizard is typically mounted by another and then lays about four eggs, which hatch about eight weeks later. This all seems pretty standard for a lizard. The main difference between these and most other species is that all individuals are female. The mounting behaviour observed during the reproductive period, typically carried out by males in most species, is performed by another female and seems to stimulate ovulation. And the four eggs all develop into females, identical (genetically) to the lizard who laid them.

There are just under 50 species of North American whiptail lizards, mainly occupying the drier regions of the southern United States and ranging into Central America. Surprisingly, more than a dozen of these have populations that consist only of females. No

Figure 2.6 *About one-third of all whiptail lizard species are parthenogenetic—females produce female offspring that are clones of themselves. This New Mexico whiptail lizard is one of these parthenogenetic females and is relatively common in the Southern United States. Why did some species like this return to asexual breeding? Or perhaps a better question is, why don't more species breed asexually?*
Credit: Greg Schechter/CC BY 2.0.

males have ever been found, and females happily and successfully produce female offspring that are clones of themselves. How did these species come to be? They generally resulted from a hybrid cross between males and females of two different sexual species. Crosses between different species typically don't work out very well; the offspring are weak and die, or sometimes develop but are usually sterile (think of the mule—a cross between a horse and a donkey). However, many whiptail lizards are different—the males fail to develop, and the females do just fine. What's more, the females don't need the males and can produce young effectively on their own.

Parthenogenesis can be obligate, in which the species reproduces only by parthenogenic means, or facultative, in which the species often reproduces parthenogenetically but sometimes sexually. Whether it be obligate or facultative, there are two ways that parthenogenesis can work. First, it can be 'apomictic', meaning that meiosis is generally left out of the process altogether. In Chapter 3, I review meiosis, but recall from your basic Biology that it is a process in sexual organisms that halves the genetic material. In humans, meiosis results in the formation of egg and sperm cells that only have a single set of chromosomes. The regular number of chromosomes (two sets) is re-established during fertilization. In apomictic organisms, the female egg already has two sets of chromosomes (is diploid), and reproduction proceeds through mitosis. For this reason, all resulting offspring are genetically identical to each other and their mother. Because meiosis is not required and is usually only compatible with pairs of chromosomes, polyploidy (3N, 4N) is not uncommon in apomictic parthenogens. This is what we find in many of the whiptail lizards.

Second, parthenogenesis can be 'automictic', where two haploid nuclei from the same female fuse and restore diploidy. In such cases, meiosis and the variation that goes along with it occurs, and the offspring will be genetically unique, having a different mixture of alleles from each other and their parent. Automixis leads to increased homozygosity and decreased variation in subsequent generations. So, the final message is that parthenogenesis produces offspring by females that can result in identical or nonidentical progeny. If this seems confusing, it is. But don't worry about it—just remember that it's a way of females giving rise to females without the genetic material of males.

Parthenogenesis is most commonly seen in invertebrates. Bdelloid rotifers and at least one species of velvet worms are obligately parthenogenetic, and many species, such as various kinds of aphids, are cyclically parthenogenetic, alternating between asexual and sexual stages during a single year. I'll talk more about them later as they have been used to suggest when it is, and isn't, beneficial to reproduce sexually. Snails, crustaceans, arachnids, and scorpions may also employ facultative parthenogenesis as a reproductive strategy. In haplodiploid species, such as ants, bees and wasps, females have pairs of chromosomes (diploid or 2N), whereas males have only single chromosomes (haploid or 1N). Although the reproductive female requires male sperm to generate female offspring, it isn't necessary to produce males (therefore, males have no fathers). Several insects use parthenogenesis as a reproductive method.[7] Often, the presence of males influences the mode of reproduction. For example, cockroaches often lay parthenogenetic eggs when housed with other females but no males, but when males are present, they reproduce sexually. The inability to find a mate also results in parthenogenesis in stick insects.

In vertebrates, parthenogenesis has been documented in over 70 species of fish, amphibians, and reptiles (including birds), although all female populations that don't require males have only been recognized in certain reptiles (such as whiptail lizards).[8] In the other groups, the eggs of females must usually be stimulated to develop by sperm contributed by males of another species. In some instances, males provide sperm, which plays no role in fertilization, while in others, the male sperm is used in fertilization but is not used during zygote formation; thus, the female genetic lineage remains intact. All of these species likely arose from hybridization between two closely related species.

In many cases, females of normally sexually reproducing species will produce offspring without interacting with a male, a situation observed relatively frequently in zoos. Perhaps they just get tired of waiting. For example, the rather fierce-sounding Asian water dragons (*Physignathus cocincinus*) are green lizards native to Southern Asia and Australia. They are often kept as pets, so they can be 'friendly' enough to be handled and held. Although it isn't unusual for females to produce infertile eggs in zoos when there are no males around, one of these eggs hatched into a female at the Smithsonian National Zoo. Remarkably, she had been kept without males since before she was reproductively mature. Investigators completed a genetic test on the offspring to confirm that they arose parthenogenetically rather than being sired by a mystery male who may have somehow snuck into the cage. Sure enough, the mother proved to be the only parent.[9] No cases of parthenogenesis have been documented in mammals.

The advantage of asexual reproduction is made clear by the success of a particular grasshopper in Australia. The more commonly called the matchstick grasshopper (*Warramaba virgo*; Figure 2.7) is probably the result of mating between a male and female of two closely related species, *W. whitei* and *W. flavolineata*, that happened about 250,000 years ago.[10] As I mentioned, crosses between different species don't usually occur, and when they do, the resulting offspring typically don't survive long or can't breed successfully. However, the cross resulting in the matchstick grasshopper produced females that could successfully reproduce without males. The remarkable thing is that these grasshoppers, thought to have arisen once, between a single cross of two sexual species, have spread across the country, expanding their range further than either of their parent species. Thus, they seem to have avoided many of the costs of sexual reproduction and thrived on the benefits of asexual reproduction. So that brings us back to why more animals don't employ asexual reproduction as a breeding method.

Parthenogenesis is a type of reproduction that doesn't require the egg to be fertilized by the male, so at first glance, it seems to fit our definition of asexual reproduction. And it is . . . in a way. On the one hand, the male provides no genetic contribution in producing offspring. Indeed, in some cases, as in whiptail lizards, there are no males anyway. Thus, reproduction is asexual, or at least 'unisexual'. On the other hand, parthenogenesis has much more in common with sexual reproduction, and most, if not all, forms have descended from sexually reproducing species. Lumping it in with fragmentation

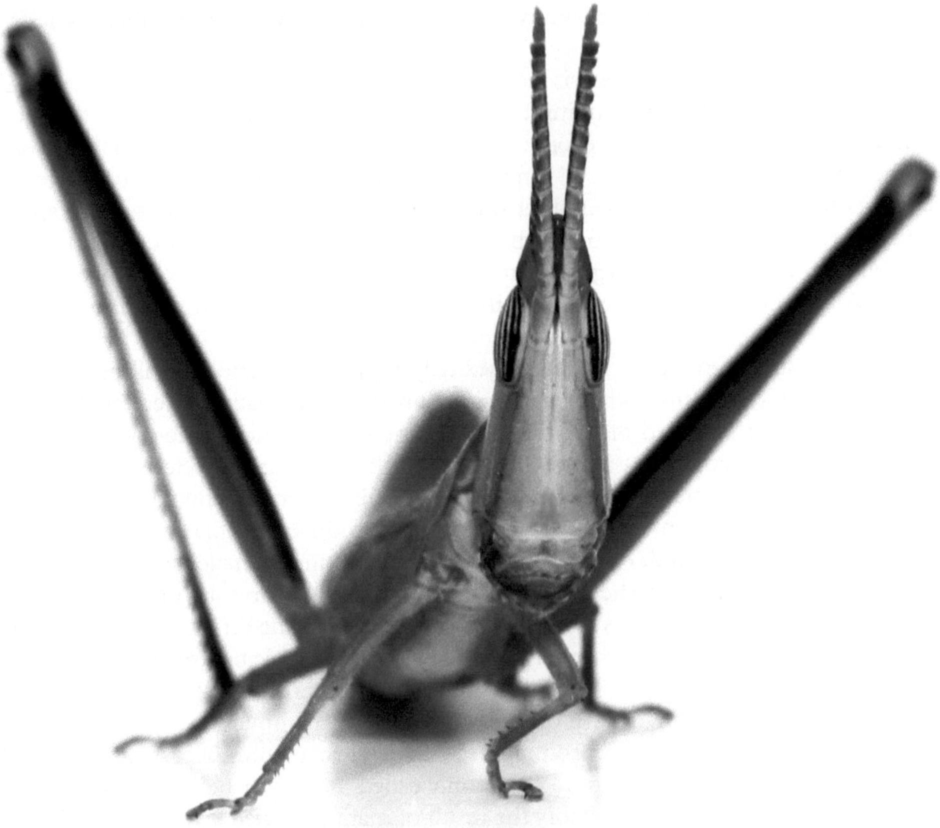

Figure 2.7 *This matchstick grasshopper illustrates the advantage that a species that has returned to an asexual breeding mode can have. Consisting of all female populations, this grasshopper has successfully expanded its range across much of Australia.*
Credit: Michael Kearney.

and budding, with which it shares few characteristics, is a bit misleading. However, put in either group at your pleasure; the main thing is that you understand what it is.

2.6 Summary

Asexual reproduction has existed since life first originated on the planet and was the only form of reproduction for almost two billion years. It remains the only type of reproduction in bacteria and archaeans and has reappeared in many eukaryotic plants and animals that were once sexual. It has the advantage of requiring only one parent;

thus, asexual individuals can contribute their full complement of genes to each of their offspring. There is no need to spend time and energy finding, courting, and copulating with other members of the population, and individuals don't have to reduce their genetic contribution to progeny by two. Common types of asexual reproduction include fission, budding, fragmentation, spores, and various vegetative techniques. However, natural selection cannot work if all individuals are the same and, thus, is important if environmental conditions change. The asexual bacteria can increase their variability by transferring genetic information in at least three ways—transformation, transduction, and conjugation. Later, transformation will be discussed in more detail as a possible evolutionary precursor to meiosis in sexual reproducers. Asexual reproduction also occurs by cells within all multicellular animals by mitosis, an essential consideration in human cancer studies. Although not discussed, many other diseases can be better understood by examining them through an evolutionary lens. Take this as a plea for all students entering into health fields to have a strong background in this field! Finally, parthenogenesis is a reproductive mode where the genotype of only one individual is transferred to the progeny, so it fits the definition of 'asexual reproduction'. However, it is clear that parthenogenesis evolved from sexual reproduction, and to include it with other forms of asexual reproduction may be misleading. As such, parthenogenetic organisms have played a valuable role in better understanding the evolution of sex, and they will reappear in the following pages.

References

1. World malaria report 2022. Geneva: World Health Organization; 2022. Licence: CC BY-NC-SA 3.0 IGO.
2. Gibson TM, Shih PM, Cumming VM, Fischer WW, Crockford PW, Hodgskiss MS, et al. Precise age of *Bangiomorpha pubescens* dates the origin of eukaryotic photosynthesis. Geology [Internet]. 2017 Dec [cited 2022 Sep 3];46(2):135–138. Available from: https://doi.org/10.1130/G39829.1
3. Boddy L. Interactions with humans and other animals. In: Watkinson SC, Boddy L, Money NP editors [Internet]. The fungi. 3rd ed. Academic Press; 2016 [cited 2022 Sept 3]. pp. 337–360. Available from: https://www.sciencedirect.com/book/9780123820341/the-fungi
4. Denning DW, O'Driscoll BR, Hogaboam CM, Bowyer P, Niven RM. The link between fungi and severe asthma: A summary of the evidence. Eur Respir J [Internet]. 2006 Mar [cited 2022 Sept 3];27(3):615–626. Available from: https://doi.org/10.1183/09031936.06.00074705
5. Zhao Z-M, Zhao B, Bai Y, Iamarino A, Gaffney SG, Schlessinger J, et al. Early and multiple origins of metastatic lineages within primary tumors. PNAS [Internet]. 2016 Feb [cited 2022 Dec 12];113(8):2140–2145. Available from: https://doi.org/10.1073/pnas.1525677113
6. Clegg SM, Degnan SM, Kikkawa J, Moritz C, Estoup A, Owens IP. Genetic consequences of sequential founder events by an island-colonizing bird. PNAS [Internet]. 2002 May [cited 2002 Dec 12];99(12):8127–8132. Available from: https://doi.org/10.1073/pnas.102583399
7. Vershinina AO, Kuznetsova VG. Parthenogenesis in Hexapoda: Entognatha and non-holometabolous insects. J Zool Syst Evol Res [Internet]. 2016 Jul [cited on 2022 Sep 3];54:257–268. Available from: https://doi.org/10.1111/jzs.12183

8. Sinclair EA, Pramuk JB, Bezy RL, Crandall KA, Sites JW Jr. DNA evidence for nonhybrid origins of parthenogenesis in natural populations of vertebrates. Evolution [Internet]. 2010 May [cited Sep 4];64(5):1346–1357. Available from: https://doi.org/10.1111/j.1558-5646.2009.00893.x
9. Miller KL, Rico SC, Muletz-Wolz CR, Campana MG, McInerney N, Augustine L, et al. Parthenogenesis in a captive Asian water dragon (*Physignathus cocincinus*) identified with novel microsatellites. PLoS ONE [Internet]. 2019 June [cited 2022 Sep 15];14(6):e0217489. Available from: https://doi.org/10.1371/journal.pone.0217489
10. Kearney M, Hoffmann A. This Australian grasshopper gave up sex 250,000 years ago and it's doing fine. The Conversation [Internet]. 2022 June [cited 2022 Sep 15]; Available from: https://findanexpert.unimelb.edu.au/news/46016-this-australian-grasshopper-gave-up-sex-250-000-years-ago-and-it%27s-doing-fine

3

The Road to Sexual Reproduction

3.1 The evolution of eukaryotes

Nobody knows for sure how life first arose. The emergence of life from nonlife has attracted experiments, computational models, religious and scientific debate, and much speculation. How it started, however, isn't our concern. Our story begins much later when the early, single-celled organisms formed unions before passing on their genetic material to the next generation. Prior to this, all life forms were asexual; they didn't require another organism to reproduce. Current estimates put the age of the Earth at about 4.5 billion years. Life appeared not long after, probably around 3.8 billion years ago, and organisms reproduced only asexually for over a billion years. While this timespan might be difficult for us to conceive, it is evident that sexual reproduction wasn't necessary for the continuation of life. So, why did sexual reproduction originate? And how? Those are the questions in which we're most interested. Most agree that the evolution of eukaryotes from their prokaryote ancestors signalled a new way of life in several ways, including reproduction. So, let's review.

As indicated in the first chapter, life on Earth is divided into two superkingdoms—Prokaryota and Eukaryota. The prokaryotes, in turn, have two kingdoms, Archaea and Bacteria. At one point, not long ago, these were both considered as one kingdom. However, they are pretty different in many ways. Many archaeans are extremophiles, meaning they can live in harsh habitats like hot springs, thermal vents, salt lakes, or the guts of ruminants. Bacteria can also produce asexual spores, which is not an archaean reproductive trait. The main differences between prokaryotes and eukaryotes are that (i) most of the genetic material of eukaryotic cells is enclosed in a membrane-bound nucleus, while that of prokaryotes floats freely in the cytoplasm; (ii) eukaryotic cells may have many other membrane-bound cytoplasmic organelles, like mitochondria and chloroplasts, that prokaryotes don't; and (iii) most eukaryotes reproduce sexually, or at least descend from species that reproduced sexually, while prokaryotes reproduce asexually. But the division of life into two major groups doesn't tell us much about the evolution of these early forms, which is what is most interesting to us.

Prokaryotes were, and are, extremely successful. For example, there are about the same number of bacterial individuals in one gram of dental plaque as all the humans

The Evolution of Sex. Kevin Lee Teather, Oxford University Press. © Kevin Lee Teather (2024).
DOI: 10.1093/9780191994418.003.0003

that ever existed.[1] They have been able to exploit a vast range of habitats and exist just about everywhere on Earth. But two things that prokaryotes (and eukaryotes) need are carbon and energy. Different species of prokaryotes obtain their carbon from different sources—from gaseous carbon dioxide or other organisms in the form of organic carbon. But how they use energy is particularly important to us. Respiration refers to the production of energy the cell can use in metabolic activities, generally in the form of adenosine triphosphate or ATP, metabolized from glucose and other sugars consumed as food. Certain species use oxygen in this chemical process, producing ATP aerobically, while others don't use oxygen and manufacture ATP anaerobically. Still others can do both. The important thing is that aerobic metabolism is much more efficient than anaerobic metabolism in producing ATP. In fact, about 16 times as many ATP molecules can be made aerobically using the same amount of glucose.

The evolution of eukaryotes from prokaryotes almost certainly involved a symbiotic relationship between an anaerobic and aerobic bacterium.[2] Oxygen increased significantly in the atmosphere and the upper layers of the ocean about 2.4 billion years ago, becoming known as the Great Oxygen Event, or GOE. This rise in oxygen levels was primarily due to cyanobacteria releasing oxygen through photosynthesis. As the theory goes, now widely accepted by almost all evolutionary biologists, a larger anaerobic archaean somehow internalized a smaller aerobic bacterium by digestion or parasitism. The smaller organism then became the mitochondria of present-day cells, responsible for aerobic energy production (Figure 3.1). Similarly, the chloroplast was a type of photosynthetic bacterium that could use the energy from sunlight to drive certain reactions by the organism; it was internalized later in the lineage that would become plants. A closer look at these organelles strongly supports the idea that they were once independent prokaryotes. First, both mitochondria and chloroplasts are enclosed by two membranes: an inner membrane that appears to have come from prokaryotes and an outer one similar to that of living eukaryotes. Second, they can multiply asexually using binary fission, a reproductive mode used by many prokaryotes. Third, and most importantly, they have their own functional genomes. The interaction between the mitochondrial and nuclear genes may dramatically influence sexual reproduction, as will become clear.

Although the evolution of the eukaryotes from the prokaryotes was a major transition that dramatically influenced life on Earth, there is a great deal we don't know. The eukaryotes are much more complex than prokaryotes, creating a gulf between them that is difficult to explain. They are, on average, about 1000 times as large as prokaryotic cells, although there are eukaryotes with smaller cells and prokaryotes with larger cells.[3] Increased complexity of the eukaryotes arises primarily from compartmentalization, endomembranes, and cytoskeleton components, which are all missing in prokaryotes. Notably, the efficiency of energy production by mitochondria, found in all eukaryotes, allows them to grow larger and be more active. What interests us most is that their reproductive mode changes from asexual to sexual, with the accompanying genetic exchange changing from lateral (between individuals of the same generation) to mainly vertical (between parents and offspring).

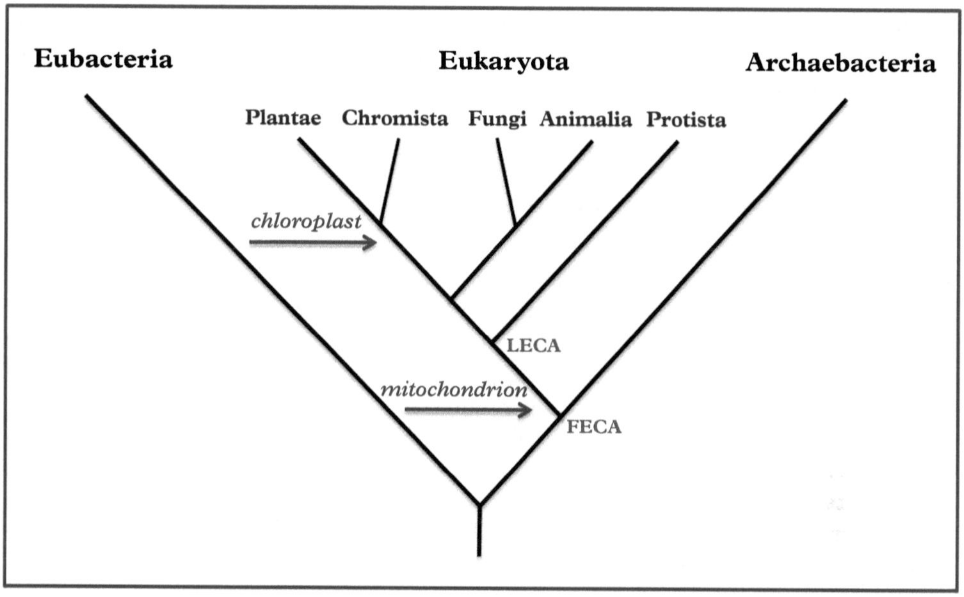

Figure 3.1 *The relationship between the three kingdoms of life. Mitochondria are thought to have originated between the first and last eukaryotic common ancestors (FECA and LECA), while chloroplasts were incorporated later. Both of these cell organelles arose from eubacteria.*

What is debated is where many of these changes occurred between the first and last eukaryotic common ancestor (FECA and LECA). As expected, eukaryotes today contain a mixture of genes that are assumed to have evolved very early, as well as many unique ones that came along later.[4] Not surprisingly, a fundamental understanding of the last common eukaryotic ancestor (LECA) has increased considerably with the recent development of modern molecular and genetic techniques. We assume that if a gene (or protein) occurs in many widely divergent, extant groups of eukaryotes, it was likely present in their common ancestor. Thus, we can reconstruct a partial gene complement of LECA by looking at present-day genes found in widely separated eukaryotic lineages.

The evolution of eukaryotes from their prokaryotic ancestors was a huge step and opened up a large number of ecological opportunities. Some would remain unicellular, like protozoans, chromists, and some fungi, while others would become multicellular (other fungi, plants, and animals). Being multicellular allowed them to partition various functions into different cells and paved the way for the development of organs and organ systems. Most importantly, the reproductive mode was to change from providing descendants with a copy of their entire genome to diluting the genes they provided to offspring by mixing them with those of another individual.

The last eukaryotic common ancestor, the organism that gave rise to all other eukaryotes, is thought to have been a metabolically sophisticated, sexually reproducing, flagellated amoeboid. It had many characteristics observed in modern groups, including the ability to reproduce sexually. Importantly, it had mitochondria, which could prove to be important to our story.

3.2 The mitochondrial genome

The genetic system of the mitochondrion is what we are most interested in, as it may have been influential in the evolution of sexual reproduction. I say 'may have been' as one of the hypotheses proposes that high-quality mitochondrial DNA may be vital when selecting appropriate mates. Not surprisingly, given its prokaryotic ancestry, the genes of the mitochondria are found in a circular ring instead of having a linear arrangement, as seen on chromosomes of the nucleus. Initially, this organelle had a lot of genes, enough to construct and run an entire organism. However, as time passed, more and more of these genes were transferred to the chromosomes in the nucleus, so today's mitochondria have only a tiny percentage of their original complement. The number of genes left in the mitochondria varies between eukaryotes, ranging from three (*Plasmodium vivax*) to almost 100 (*Reclinomonas americana*).[5] In general, the size of the mitochondrial genome is only about 0.5% of that used to drive its bacterial ancestor. While most mitochondrial genes have been lost, many others have been transferred to the nuclear chromosomes. Indeed, most of the genes required for the proper functioning and reproduction of the mitochondria actually reside in the nucleus. Still, the genes that remain are involved in a critical process.

Since the mitochondria are involved in aerobic energy production, you wouldn't be surprised to discover that its genes code for products used in this process. Specifically, in animals, the mitochondria have about 13 genes that code for proteins that all function in oxidative phosphorylation (OXPHOS), the series of biochemical reactions that produce ATP, the cell's primary energy source (Figure 3.2). The mammalian mitochondrion also contains 24 other genes used in the translation of mitochondrial proteins (22) or related to mitochondrial ribosomes (2). Notably, the mitochondrial-produced proteins involved in oxidative phosphorylation work closely with proteins produced by the nucleus. That isn't to say that the mitochondrion isn't required for other processes (e.g. normal cell signalling and programmed cell death). However, the genetic material that directs these processes resides in the nucleus, and besides, these functions aren't that important in any discussion about sexual strategies.

A few things that are necessary to understand before we go any further. First, the genetic material in the mitochondria has a relatively high mutation rate. In invertebrates, mutation rates in the DNA of mitochondria are about 2 to 6 times faster than in the nucleus, while in vertebrates, they're about 20 times faster.[6] This high mutation rate means that the hundreds of mitochondria in a typical cell (red blood cells don't have any mitochondria, while liver cells may have 2000) may have slightly different DNA. Second, there is generally no crossing-over between chromosomes since the division of mitochondria only involves mitosis (see Section 3.3 for an overview of meiosis and

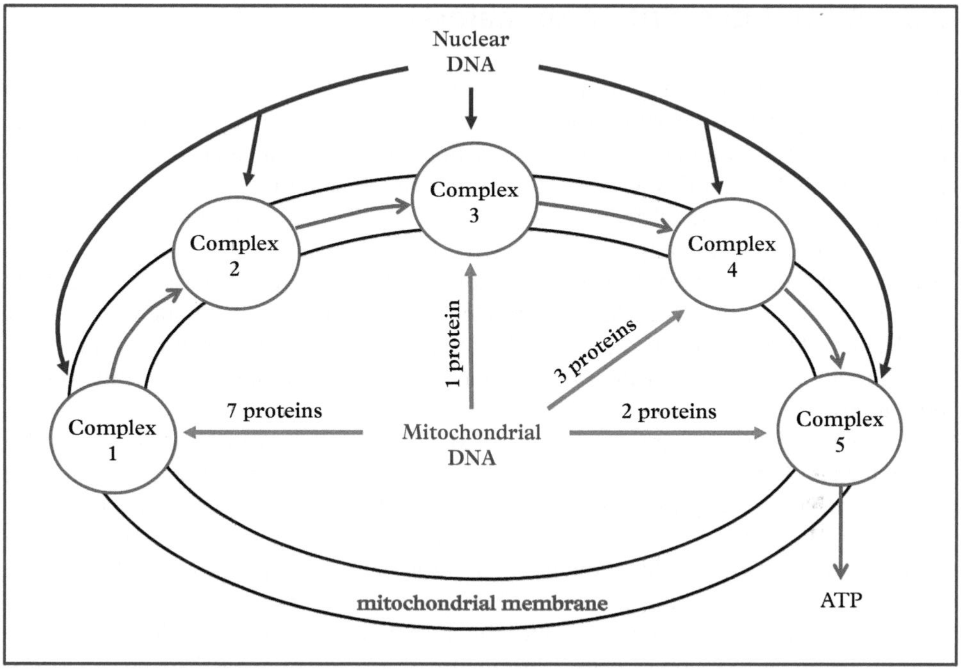

Figure 3.2 *Gene interactions between the nuclear and mitochondrial chromosomes are responsible for oxidative phosphorylation and the production of ATP. Mitochondrial DNA codes for thirteen proteins used in this process. These interact with nuclear gene products in five complexes that occur on the mitochondrial membrane.*

crossing-over). Later in this chapter, I discuss why crossing-over adds to the variability in sexually produced offspring. As a result, the 'offspring' of mitochondria are largely clonal versions of their parents, with any differences between mitochondria within a cell resulting from high mutation rates. Third, the DNA of the mitochondrion is almost always maternally inherited. There are mitochondria in sperm that are needed to provide energy for its voyage to the egg, but these either break down just after fertilization or never enter the egg in the first place.

Therefore, the cell needs two sets of genes to provide the necessary structures for an efficient energy-producing mechanism—one set from the mitochondria and the other from the nucleus. The genes in the mitochondria are always handed down from the mother and are subject to high mutation rates, making them somewhat variable. The genes present in the nucleus are the product of sexual reproduction between two individuals in eukaryotic organisms. In other words, they are derived from both the mother and father. The interactions between the genes of the mitochondrion and the nucleus are referred to as 'mitonuclear' interactions. Don't be too concerned if the process of oxidative phosphorylation necessary for aerobic respiration seems far removed from the evolution of sex and sexual strategies—this will become clearer as we continue our story.

3.3 A brief description of meiosis

Two processes had to develop during the evolution of eukaryotes from prokaryotes that would broadly define sexual reproduction. First, two cells had to recognize each other and fuse; this resulted in a doubling of genetic material. Organisms, however, couldn't permit this doubling of DNA to occur every generation, and the second process involved halving the genetic material again before the next round of reproduction. These were the processes of fertilization and meiosis, which essentially distinguish sexual reproduction from asexual reproduction. In the next chapter, I examine the evolution of the gametes and briefly discuss how gametes are attracted to each other and ultimately fuse during fertilization. Here, I look at meiosis and attempt to make clear that understanding the role of meiosis provides us with a better understanding of why sexual reproduction is so ubiquitous.

All sexually reproducing organisms alternate between diploid and haploid stages. The diploid stage dominates in animals, but this isn't true for many other eukaryotes. For example, what we recognize as mosses are actually the haploid stages in their life cycles. But let's stick to animals for the moment. All individuals have numerous pairs of chromosomes, one of the pair from the father and the other from the mother. Humans, for example, have 46 chromosomes, with 23 donated by each parent. These identical sets of paired chromosomes (except for ones that determine sex) are found in most types of cells in the body. Division of cells, beginning with the initial one-celled zygote, occurs by mitosis so that each cell is genetically identical to its parent (except for mutations). In other words, mitosis produces clones and is used to reproduce nearly all cells in a diploid body; a similar process occurs in asexual reproduction. Differences in cell types are brought about by which genes are turned on and off, not by any differences in the genes found in different cells.

Just before meiosis begins (Interphase), each homologous chromosome replicates so that it consists of two sister chromatids joined together by a centromere. Think of this as preparation for meiosis. While I won't describe it fully here, Table 3.1 provides an overview of meiosis, including two stages each for prophase, metaphase, anaphase, and telophase. These are shown in Figure 3.3. Meiosis I and II are similar but have a couple of essential differences. Meiosis I starts with one diploid cell whose genetic material has been duplicated, while Meiosis II begins with two diploid cells and ends with four haploid cells. Besides the independent assortment of alleles, the event that we are most interested in occurs near the beginning of Meiosis I, crossing-over, and thus, I'll concentrate on this.

During Prophase I, the homologous chromosomes, each consisting of a pair of sister chromatids, come together in a process called synapsis and are called tetrads (because they consist of four sister chromatids). Although the chromosomes, and thus the genes, from both parents align, the two homologues may have different alleles (versions) of those genes. For example, the chromosome inherited from the father might have the allele for brown eyes, while that coming from the mother has the allele for blue eyes. Crossing-over allows similar regions (containing the same or different alleles) to be transferred from one homologue to the other (Figure 3.4). The two homologues join

Table 3.1 *The various stages of meiosis and their main characteristics.*

Interphase		**DNA replication; duplicated chromosomes are called sister chromatids and are joined together by a centromere**
Meiosis I	*Prophase I*	Sister chromosomes condense and become visible under light microscope; homologous chromosomes synapse; recombination takes place
	Metaphase I	Homologous pairs align along equator of cell; each pair joined by microtubules to centromeres on opposite sides of cell
	Anaphase I	Homologous chromosomes, consisting of sister chromatids, pulled to opposite sides of the cell
	Telophase I	New nuclear membrane forms around each pair of sister chromatids; cytoplasmic membrane (or wall) may form between two daughter cells; each cell is now haploid
Meiosis II	*Prophase II*	Disappearance of nuclear membrane; spindles begin to form
	Metaphase II	Sister chromosomes align along mid region of cell and attached by spindles to centrosomes on opposite sides
	Anaphase II	Sister chromatids pulled to opposite sides of cell
	Telophase II	Chromosomes become more diffuse; nuclear membrane reforms; cell membrane forms

at chiasmata (sing. chiasma), where this crossing-over occurs. Therefore, a chromosome the mother donated potentially ends up with some paternal alleles on it, and vice versa.

Recombination of genes results in variable offspring and is one reason individual progeny are not identical to their parents or siblings. However, the number crossing-over sites between homologous chromosomes is limited; it was estimated that, among fifty eukaryotes, about 80% of homologous chromosomes had fewer than three cross-over events.[7] In addition, although the points at which crossing-over occur are often viewed as randomly determined (facilitating the mapping of recombination events), they are not. In humans, for example, there are about 30,000 hotspots generally found outside of genes.[8] However, these hotspots tend to be spaced relatively evenly, meaning that genes located further apart on chromosomes are more likely to be separated by cross-over events. Conversely, genes close together are less likely to be split up. This arrangement means the chance of alleles being split up on one chromosome, or the recombination rate, can be determined.

The remaining events that occur during meiosis can be summarized rather quickly, as they are less important to us. The newly formed pairs of chromosomes are pulled apart, and cellular and nuclear (sometimes) membranes form around each set. Thus, each new cell consists of a duplicated set of chromosomes, originally coming from one

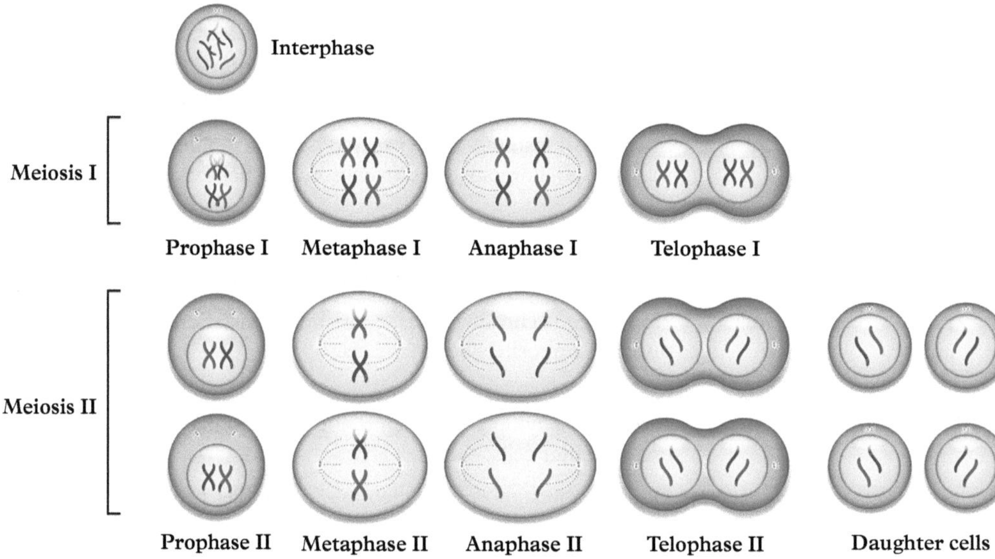

Figure 3.3 *An overview of the stages of meiosis. In this case, the diploid number is two (four chromosomes in total). The genetic material doubles before Meiosis I during Interphase (four pairs of sister chromatids). Crossing-over between homologous chromosomes (shown in more detail in Figure 3.4) occurs during Prophase I. The result of meiosis is four haploid cells originating from one diploid cell.*

Credit: olando/Adobe Stock Photo ID 273158166.

parent but with some alleles of the other depending on the number of cross-over events. This result signals the end of Meiosis I. Meiosis II is similar to mitosis. The two cells now divide, with a single set of chromosomes going to each daughter cell, resulting in four haploid cells containing a single copy of the chromosomes.

To summarize, meiosis typically results in cells that (i) contain half the DNA of their parent and (ii) are all genetically unique. During meiosis, one cell produces four offspring cells, each containing a single (haploid), not double (diploid), set of chromosomes. The haploid products of humans and most other multicellular eukaryotes are unicellular sperm (or pollen in plants) and eggs. Every haploid gamete gets a unique set of genes; this is made possible by the independent assortment of the chromosomes as they migrate into different gametes and because of the production of new gene combinations resulting from genetic exchange between the paternal and maternal (homologous) chromosomes. This crossing-over results in recombination or new gene combinations on a single chromosome.

Meiosis is integral to sexual reproduction, so you should understand the main processes before going further. The variability that meiosis creates is an essential feature of sexual reproduction, and I'll refer to it repeatedly in the coming chapters. But

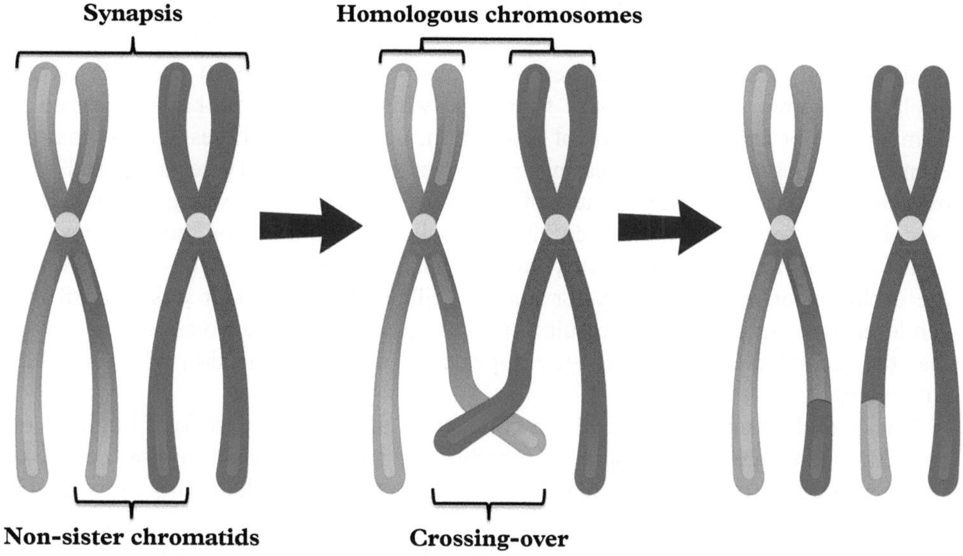

Figure 3.4 *Crossing-over during Prophase I of meiosis occurs between non-sister chromatids of homologous chromosomes. This process results in the formation of recombinant chromatids. Sister chromatids separate in Anaphase II, and haploid cells at the end of meiosis contain unique genetic instructions.*

Credit: VectorMine/Adobe Stock Photo ID 345855676.

was this the selection pressure that led to its evolution? We can only speculate as it happened during the transition between prokaryotes and eukaryotes about 1.5 billion years ago, but I'll address two hypotheses concerning the evolution of meiosis now.

3.4 The evolution of meiosis

The evolution of meiosis has been challenging to explain because it involves complexity in functions necessary for all sexually reproducing organisms. It encompasses two stages, Meiosis I and II, and although similar in some ways to mitosis, several processes make it unique. Meiosis probably evolved during the transition between the first and last common ancestors of the eukaryotes (see Figure 3.1). We are relatively confident that meiosis was present in LECA, as it is found in all eukaryotic groups today. But where did it come from, and what factors stimulated its evolution? The first part of that question is mechanistic (proximate), while the second, and more interesting (to me at least), is functional (ultimate).

3.4.1 The repair of DNA: Meiosis from transformation

In the period after life evolved, aerobic prokaryotes made use of the little oxygen available to power their metabolic functions, while anaerobic prokaryotes relied on fermentation or anaerobic means of respiration. Current hypotheses suggest that two events were closely tied to increasing oxygen levels. First, eukaryotes appeared when oxygen increased sufficiently because of the activity of photosynthesizing bacteria. However, because the efficiency of aerobic metabolism is required for the complexity of organisms to increase substantially, the still relatively low level of oxygen prevented the evolution of multicellularity for another billion years or so. It wasn't until oxygen levels rose sufficiently as a result of photosynthesizing prokaryotes and eukaryotes that the second event occurred—the appearance and diversification of multicellular eukaryotes.

Davies and Ursini[9] introduced the concept of the 'oxygen paradox', pointing out that although oxygen is necessary for a highly efficient metabolism in most cells, oxygen is also dangerous to these same cells. So, what's the problem with oxygen? The genetic material of cells is continually bombarded by substances and processes that may be harmful, including UV and ionizing radiation, mutagenic chemicals, viruses, and, importantly, oxidative processes resulting from cellular metabolism (Figure 3.5). Although oxygen can be used much more efficiently to produce ATP, the primary energy source for the cell, the resulting damage it can cause to DNA is significant. Aerobic metabolic activities of the cell release reactive oxygen species and other harmful products; oxygen radicals are a particularly serious concern for the integrity of DNA as they readily react with the DNA itself. In fact, reactive oxygen species are the most serious problem for DNA in the mammals studied. In rats, for example, about two-thirds of the 130,000 DNA injuries per cell per day are due to oxidative damage.[10]

DNA damage is a significant problem for all living organisms. It can be defined as a modification to DNA that can affect its ability to engage in normal replication or translation. It may include (but is not limited to) insertions or deletions of nucleotides and breaks in one or both strands of the double helix. This damage isn't a problem in most cases, as numerous repair mechanisms are available to fix the impaired DNA. In fact, all organisms invest heavily in such repair mechanisms, suggesting that the problem can be quite serious. In humans, for example, the products of more than 125 genes are involved in the repair of DNA.[11] The problem is that these processes don't always work, and DNA damage can build up in cells, resulting in many diseases and aging. One serious problem is a break in the DNA strand. While single-stranded breaks can use the complimentary strand in a double helix, double-stranded breaks may interfere with replication and lead to the cell's death. The most efficient way of fixing them is using an undamaged homologous chromosome (or piece of chromosome) as a template.

Homologous recombination, to fix double-stranded breaks in the DNA (and other DNA damage), requires a homologous, undamaged chromosome, which can either come from a sister chromatid (formed in preparation for mitosis) or another, usually closely related, individual. Unfortunately, mitotic recombination isn't particularly efficient, and homologous DNA from this process is only available briefly during the cell

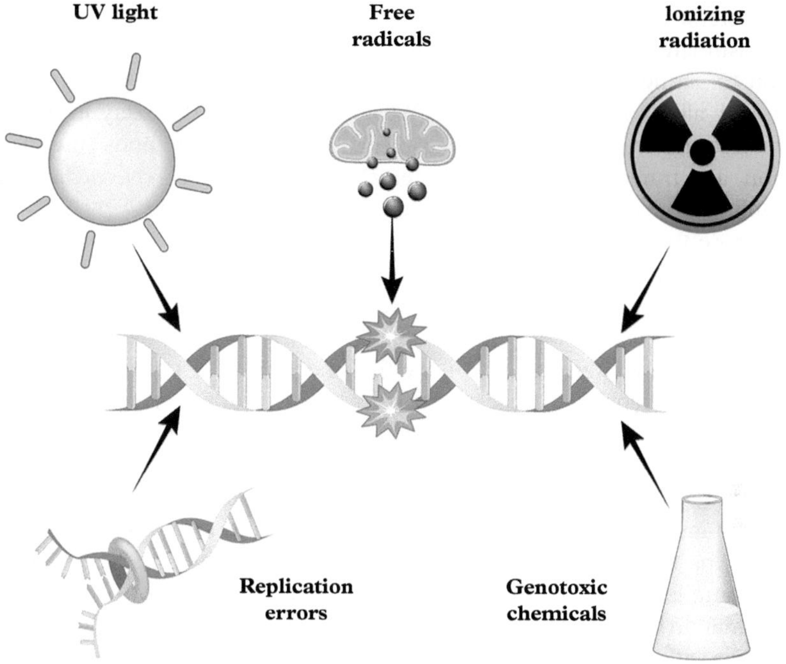

Figure 3.5 *There are many processes or events that can result in damage to the genetic material. In eukaryotes, oxidative phosphorylation, carried out by the newly acquired mitochondria, produces free oxygen radicals that could harm DNA. With aerobic metabolism increasing as the level of oxygen in the air increased, repairing this damage was likely crucial in the evolution of the eukaryotes.*
Credit: designua/Adobe Stock Photo ID 454418728.

cycle. Therefore, foreign DNA is often used. This brings us back to transformation in prokaryotes.

Transformation, which involves the uptake of DNA from one prokaryote into another, was mentioned briefly in the last chapter. After a prokaryote dies, its DNA is released into the environment, where it often floats as linear stretches of a double DNA helix. Before taking in any of this DNA, another prokaryote must be competent. In other words, physiological changes occur in the prokaryote to prepare it for taking up the extracellular DNA. The fact that competency is more likely to occur when the organism is stressed is not a coincidence; the DNA of the recipient cell is most likely damaged under such conditions and needs to be repaired. Competency involves the production of proteins required to (i) transfer the extracellular DNA across the cell membrane and (ii) incorporate the new genetic material into the circular chromosome. As the DNA enters the cell, one strand degrades while the remaining strand can attach to a homologous region of the circular chromosome. The process is random, but in some cases, the extracellular DNA will replace a damaged portion in the second prokaryote.

Many bacteria across diverse taxonomic groups today engage in transformation, and recent estimates confirm more than 80 species undergo transformation naturally.[12] In addition to acting as a repair mechanism for damaged DNA, transformation can introduce new alleles to the prokaryote. We see this frequently today in disease resistance, which results from any of the three lateral gene transfer methods (transformation, conjugation, and transduction; see Chapter 2). For example, *Streptococcus pneumoniae* is commonly found in the respiratory tract of humans. Specific forms, or serotypes (there are more than 100), cause or contribute to a wide range of diseases, including tuberculosis, conjunctivitis, otitis media, sinusitis, and acute bronchitis. Resistance to antimicrobials and vaccines is known to result from acquiring resistant alleles through transformation.[13]

In both meiosis and transformation, we find the pairing of chromosomes from two different cells, an exchange of alleles between these chromosomes, and a transfer of this new genetic arrangement to daughter cells (Figure 3.6). Most importantly, both processes use many of the same proteins to accomplish these events. The RecA protein, for example, that forms during competency in *Streptococcus pneumoniae* and other prokaryotes is instrumental in catalysing the recombination of two homologous chromosomes during transformation.[14] Two proteins closely related to RecA—Rad51 and Dmc1—accomplish the same thing in meiosis. The relatedness of proteins necessary for transformation and meiosis suggests that they rely on the same genetic material to complete their activities.

The prokaryotic ancestor of eukaryotes was probably capable of transformation, and meiosis may have evolved from this process as a repair mechanism for DNA damage, much of which was caused by reactive oxygen species. So, meiosis was present early in the lineage of eukaryotes and was maintained in almost all modern forms. In addition to repairing damage necessary in the 'immortal' DNA of the sex cells, it was also involved in converting diploid to haploid cells. It was once thought that some of the earliest forms of eukaryotes were asexual and thus had no need of meiosis. For example, *Giardia intestinalis*, *Leismanias sp.*, and *Trichamona vaginalis* are extant asexual protists initially thought to arise before meiosis evolved in eukaryotes. It is now known, however, that these species can reproduce sexually (or have many of the genes necessary for meiosis); therefore, it is believed that they have descended from sexual types.[14]

3.4.2 The improvement of recombinational accuracy: Meiosis from mitosis

The second hypothesis concerning the evolution of meiosis states that this process was derived from mitosis as detailed by Wilkins and Holliday.[15] Of course, this might tell us how it evolved, but not why. However, let's look at both the similarities and the differences between meiosis and mitosis, and this might suggest why meiosis became necessary in eukaryotes.

Table 3.2 compares mitosis and meiosis, indicating which processes are novel. These include (i) the pairing of homologous chromosomes (syngamy), (ii) recombination

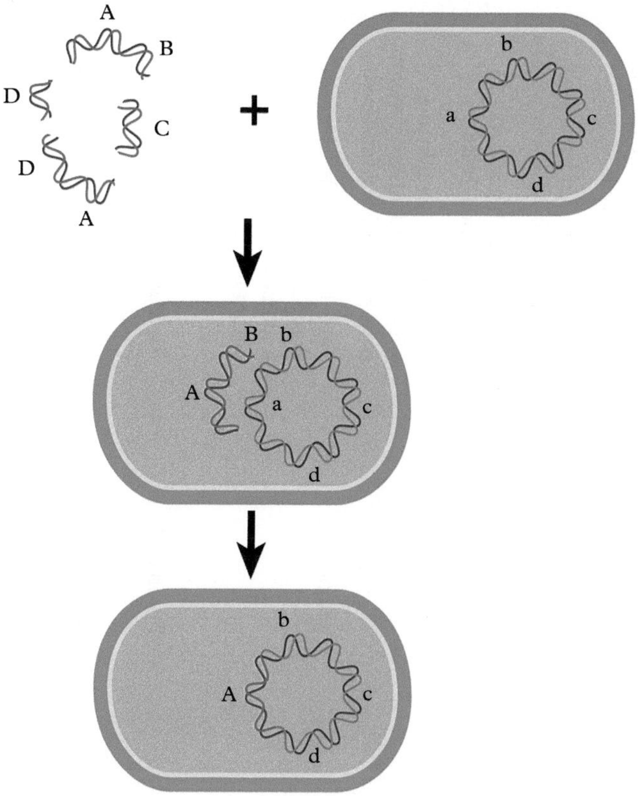

Figure 3.6 *Transformation in prokaryotes is similar to recombination during meiosis. In both cases, exchanging nuclear material between two unrelated strands of DNA can both fix errors and add different genes or alleles to the recipient's DNA. In the case of transformation, shown here, free-floating strands of DNA are taken up by a competent prokaryote and incorporated into its circular chromosome.*
Credit: Alila Medical Media/Adobe Stock Photo ID 188786409.

between homologous chromosomes, (iii) sister chromatids remaining attached (during Meiosis I), and (iv) the absence of DNA replication before the start of Meiosis II. Although there is a tendency to focus on recombination, as it is one of the hallmarks of meiosis, the ability to recombine genes is found in the prokaryotes during mitosis (although at a significantly reduced rate). Thus, the machinery for this process, as well as the last two activities, likely already existed when meiosis evolved. In fact, Wilkins and Holliday argued that the only truly novel feature of meiosis was the accurate lining up of the two homologous chromosomes before crossing-over could take place. Therefore, it's necessary to ask why it was important for homologous chromosomes to line up correctly.

There is a serious problem with recombination if the parts of DNA being aligned don't match. Crossing-over between two different parts of non-sister chromosomes can have serious consequences, even resulting in the organism's death. The alignment and

Table 3.2 *A comparison between mitosis and meiosis. The four processes in italics are those specific to meiosis. However, only homologue pairing (synapsis) requires an evolutionary explanation, as the molecular machinery already existed for the other processes. This table has been recopied15 with permission from Oxford University Press.*

Mitotic stage	Result	Meiotic stage	Result
S phase	Chromatid duplication	S phase I	Chromatid duplication; DNA breaks introduced
Prophase	Chromosome condensation	Prophase I	Chromosome condensation; *homologue pairing, recombination*
Metaphase	Chromosome alignment in centre of spindle body	Metaphase I	Alignment of homologues in centre of spindle body
Anaphase	Centromere splitting; chromatids separated	Anaphase I	Separation of homologues with independent assortment; *centromere splitting suppressed*
Telophase	Chromatid decondensation; two daughter nuclei with mother-cell ploidy, single-chromatid chromosomes	Telophase I	Partial or complete chromatid decondensation; two hap-loid nuclei with replicated chromosomes
		Prophase II	*No S phase*; chromosome condensation
		Metaphase II	Alignment of replicated chro-matids
		Anaphase II	Centromere splitting; separation of chromosomes
		Telophase II	Chromatid decondensation; four haploid nuclei, single-chromatid chromosomes

crossing-over of homologous chromosomes were unlikely to have evolved as part of a DNA repair mechanism, as such mechanisms were already present. Such processes had been selected in the prokaryotes to deal with damage under harsh environmental conditions, such as those experienced by the archaean extremophiles. The 'meiosis from mitosis' hypothesis suggests that synapsis, the only new process, was selected so that homologous chromosomes could accurately line up; the benefit would be to reduce any mistakes that were to occur during recombination. In other words, synapsis promoted accurate sequence alignment of paired chromosomes, which would result in fewer genetic errors resulting from crossing-over between nonhomologous chromosomes or parts of chromosomes. Thus, although Wilkins and Holiday admitted that eukaryotic cells would have required efficient DNA repair, they suggested that the precise lining up of homologous chromosomes from different individuals more likely evolved to reduce crossing-over errors.

It's apparent that much of the genetic machinery used during meiosis was already present to carry out both transformation and mitosis[16] and had evolved to benefit those processes. Thus, the two hypotheses regarding the mechanistic origin of meiosis are not mutually exclusive. However, they differ concerning the selection pressure(s) that promoted this integral part of sexual reproduction—one stating that it was to repair damaged DNA and the other suggesting it evolved to reduce errors during crossing-over. And what about increased variability? Such variability in offspring is a third hypothesis concerning the selective advantage of recombination. As I discuss in Chapter 6, having genetically variable offspring may be a real advantage in maintaining sex. But let's leave that for now and examine how two individuals come together to exchange genetic material in the next chapter.

3.5 Summary

Prokaryotes were the only form of life on Earth for more than a billion years, and prokaryotes reproduced asexually. The evolution of eukaryotes from prokaryotes, probably stimulated by a rise in oxygen levels 2.4 billion years ago, represented a fundamental leap in our evolutionary past. One of the most notable events during this period was the endosymbiotic relationship between an anaerobic archaean and an aerobic bacterium, which later became the mitochondria of all living cells. The genes of the mitochondria were largely lost or taken up by the nuclear chromosomes; however, a few functional ones persisted and are essential in working with the nuclear genome to regulate the energetics of the cell. The genes found in the mitochondria may be a critical component of mate selection, discussed later.

Meiosis was another process that evolved between the first and last common eukaryotic ancestors. Meiosis is often associated with the creation of genetic diversity, although this may not have been its original function. Crossing-over between homologous chromosomes during Prophase I of meiosis may result in the alleles of two individuals being exchanged; sets of modified chromosomes are then randomly transferred into sex cells or gametes. Both processes result in eggs and sperm having a different genetic makeup from each other and their parent. After a brief overview of the main steps occurring during meiosis, I presented two hypotheses that have been proposed to explain its evolution. First, meiosis initially served to repair damaged DNA and evolved from transformation, a process used to transfer and repair DNA in many prokaryotes. Second, meiosis functioned to prevent recombinational errors and evolved from mitosis, an activity with which it shares several features. Whichever is correct, meiosis was present in the earliest eukaryotes and is found in all the major groups of eukaryotes today as an integral part of reproduction.

References

1. Microbiology by numbers. Nat Rev Microbiol [Internet]. 2011 Aug [cited 2002 Sep 15];9:628. Available from: https://doi.org/10.1038/nrmicro2644
2. Sagan L. On the origin of mitosing cells. J Theor Biol. 1967 Mar;14(3):225–274.
3. Butterfield NJ. Early evolution of the eukaryote. Paleontology [Internet]. 2015 Nov [cited 2022 Sep 16];58(1):5–17. Available from: https://doi.org/10.1111/pala.12139
4. Koonin EV. The origin and early evolution of eukaryotes in the light of phylogenomics. Genome Biol [Internet]. 2010 May [cited 2022 Sep 17];11:209. Available from: https://doi.org/10.1186/gb-2010-11-5-209
5. Desmond E, Brochier-Armanet C, Forterre P, Gribaldo S. On the last common ancestor and early evolution of eukaryotes: Reconstructing the history of mitochondrial ribosomes. Res Microbiol [Internet]. 2011 Jan [cited 2022 Sep 18];162(1):53–70. Available from: https://doi.org/10.1016/j.resmic.2010.10.004
6. Allio R, Donega S, Galtier N, Nabholz B. Large variation in the ratio of mitochondrial to nuclear mutation rate across animals: Implications for genetic diversity and the use of mitochondrial DNA as a molecular marker. Mol Biol Evol [Internet]. 2017 Nov [cited 2022 Sep 18];34(11):2762–2772. Available from: https://doi.org/10.1093/molbev/msx197
7. Fernandes JB, Seguéla-Arnaud M, Larchevêque C, Lloyd AH, Mercier R. Unleashing meiotic cross-overs in hybrid plants. Proc Natl Acad Sci USA [Internet]. 2017 Nov (cited 2022 Sept 20];115(10):2431–2436. Available from: https://doi.org/10.1073/pnas.1713078114
8. Baudat F, Buard J, Grey C, Fledel-Alon A, Ober C, Przeworski M, et al. PRDM9 is a major determinant of meiotic recombination hotspots in humans and mice. Science [Internet]. 2009 Dec [cited 2022 Sep 20];327(5967)836–840. Available from: https://doi.org/10.1126/science.1183439
9. Davies KJ, Ursini F. The oxygen paradox. Padova, Italy: Cleup University Press, 1995. p. 811.
10. Bernstein C. Sex as a response to oxidative damage. In: Aruoma OI, Halliwell B, editors. DNA & free radicals: Techniques, mechanisms, & applications. London: OICA International; 1998. pp. 99–118.
11. Christmann M, Tomicic MT, Roos WP, Kaina B. Mechanisms of DNA repair: An update. Toxicology [Internet]. 2003 Nov [cited 2022 Aug 31];193(1–2):3–34. Available from: https://doi.org/10.1016/s0300-483x(03)00287-7
12. Johnston C, Martin B, Fichant G, Polard P, Claverys JP. Bacterial transformation: Distribution, shared mechanisms and divergent control. Nat Rev Microbiol [Internet]. 2014 Feb [cited 2022 Sep 1];12(3):181–196. Available from: https://doi.org/10.1038/nrmicro3199
13. Kurushima J, Campo N, van Raaphorst R, Cerckel G, Polard P, Veening, J-W. Unbiased homologous recombination during pneumococcal transformation allows for multiple

chromosomal integration events. eLife [Internet]. 2020 Sep [cited 2022 Sep 1];9:e58771. Available from: https://doi.org/10.7554/eLife.58771

14. Bernstein H, Bernstein C. Evolutionary origin and adaptive function of meiosis. In: Bernstein C, Bernstein H, editors. Meiosis [Internet]. IntechOpen. 2013 Apr [cited 2022 Sep 20]. Available from: https://doi.org/10.5772/56972 DOI 10.5772/56972
15. Wilkins AS, Holliday R. The evolution of meiosis from mitosis. Genetics [Internet]. 2009 Jan [cited 2022 Sep 15];181(1)3–12. Available from: https://doi.org/10.1534/genetics.108.099762
16. Bernstein H, Bernstein C. Evolutionary origin of recombination during meiosis. Bioscience [Internet]. 2010 Jul [cited 2022 Oct 5];60(7):498–505. Available from: https://doi.org/10.1525/bio.2010.60.7.5

4

Two Sexes

Early Days

4.1 The biological difference between males and females

If you are heterosexual, single, and looking for a partner of the opposite sex, recognizing a potential mate is the least of your problems. The sexes differ in average muscle mass, bone mass, and the amount and distribution of body fat and hair. We often accentuate these differences by how we dress and our lifestyle. And these are just some of the physical differences that are obvious to us. External genitalia are a dead giveaway, but these aren't often readily apparent. Of course, recognizing a member of the opposite sex is just the first step in the whole courtship process, but knowing the difference between males and females is crucial when reproduction is the goal.

But what defines a 'male' and a 'female'? Biologically, that is. Males and females of many species are vastly different and, therefore, easily recognized. Among mammals, the male is often bigger, may have ornaments at least part of the year, and could even be a different colour. During the mating season, males and females are often even easier to tell apart as some of these differences may become magnified. Quite often, external reproductive organs such as the male penis or female vulva of mammals are obvious, and the behaviour of each sex may differ. However, during much of the year, a closer examination may be necessary. Do you think you could recognize a male versus female collared peccary (*Pecari tajacu*; Figure 4.1)? Collared peccaries are pig-like animals, often called jacanas, that roam around the Southern United States and Central America in bands of about a dozen individuals. Male and female peccaries are the same size, and males have no ornamentation that makes them more attractive to females or to ward off rivals. Closer examination may reveal differences in external reproductive anatomy, such as the testes of males, but who wants to get that close to a wild collared peccary?

Birds are another group where male and female differences may or may not be noticeable. In some cases, males develop plumage differences during the breeding season, making them easily distinguishable from females. The bright yellow of the male

The Evolution of Sex. Kevin Lee Teather, Oxford University Press. © Kevin Lee Teather (2024).
DOI: 10.1093/9780191994418.003.0004

Figure 4.1 *In many organisms, males and females are almost impossible to distinguish by appearance, as in these collared peccaries. What distinguishes them as males and females? The only characteristic that is always different between the two sexes, in all sexually reproducing species, is the size of their gametes.*

Credit: Bernard Dupont/CC-BY-SA 2.0.

American goldfinch (*Spinus tristis*), the intense red of the male northern cardinal (*Cardinalis cardinalis*), and the many colours of the male painted bunting (*Passerina ciris*) make them distinct from females. However, what about American coots (*Fulica americana*)? In this species, males and females are almost identical. Male blue jays (*Cyanocitta cristata*) and American robins (*Turdus americanus*) are slightly larger, on average, than females, but these differences are mainly statistical and challenging to see from a distance. And for many species where males develop specialized plumage during the breeding season, the sexes don't differ at other times of the year. We can't even rely on external genitalia to identify the sexes in most birds, as males and females have a cloaca (a single opening for the digestive and reproductive tracts), which is similar in appearance regardless of whether the individual has ovaries or testes.

In other animals, such as the collared peccary, males and females may have few physical differences readily apparent to us. Among other vertebrates are many examples of fish, amphibians, and reptiles whose biological sex is difficult to determine by their external appearance. The same applies to many species of invertebrates. Even though size, plumage, colour, ornamentation, and even body shape can be used to distinguish sexes in some species reliably, the same characteristics are completely unreliable in others. Of course, odours, vocalizations, and behaviours may not be as evident to us but might be used by one sex of that species to identify potential breeding partners. But, given that identifying an individual's sex is essential, at least if they are to be involved in reproduction, we are still left with the problem of how such differences originated. And, on a broader level, why two sexes? Why not three? Or four? Or even (if evolution is all about efficiency) one? Before getting into that, however, let's discuss the main difference between males and females.

Several of you have guessed that females have ovaries, while males have testes. While this is partially correct, it doesn't work well for plants and some animals. Instead, in multicellular eukaryotes, males produce male gametes, while females manufacture female gametes (Figure 4.2). Gametes? These are just the cells that come together during the process of fertilization. In animals, we call them sperm and eggs. In plants, they are called pollen and eggs. I refer to multicellular eukaryotes for a reason despite the existence of many unicellular eukaryotes. However, as I'll soon point out, these often produce gametes that don't distinguish them as male or female. Of course, we often get male and female organs together in one organism—these are known as simultaneous

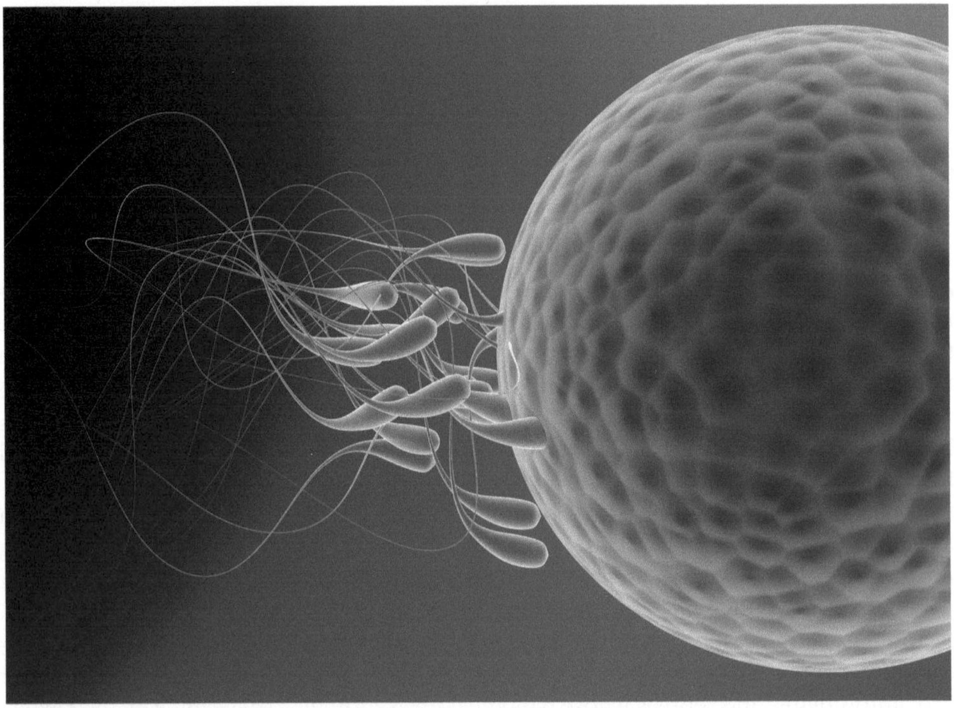

Figure 4.2 *Using similar amounts of energy to produce gametes, an individual can either make fewer large gametes or many smaller gametes. Models have also documented advantages in the two cells coming together if the smaller gamete is mobile and the larger is stationary.*
Credit: JumalaSika ltd/Adobe Photo ID 22794721.

hermaphrodites and will be examined more closely later. Regardless, there are significant differences between male and female sex cells; as we shall see, these are consistent between multicellular eukaryotes. In addition, they often lead to many other differences in behaviour and anatomy observed between the sexes. So, the difference in gametes distinguishes and defines the sexes: males produce many small, mobile gametes, while females produce fewer large, stationary gametes. Period. We only have these extremes. In other words, there is no biological sex that produces gametes of an intermediate size. I'll discuss why shortly. However, now that we have our biological definition of the two sexes, let's look at some problem areas.

4.2 Is gamete size the only way to define the sexes?

But wait. Many other ways are often used to differentiate males and females. In addition to gametes, genetics and hormones are two of the most common biological properties that most likely differ between the sexes, at least in eukaryotic animals. Males and females are also likely to differ behaviourally, in their habitat use, in many anatomical and physiological features, longevity, and a slew of other characteristics depending on the species. And there's the rub: these differences are often specific to the organisms being examined. Females of all species do not have two X sex chromosomes, although this configuration is most common in humans and many other animals. Indeed, the biological sex of an organism may not even be determined genetically. And while the hormonal regulation of sex-specific processes is relatively well-understood in vertebrates, this is not the case for invertebrates.[1] Thus, while often useful for the broadscale classification of males and females in various groups, these characteristics are not very reliable when we extend our definition to eukaryotes in general.

Humans are good examples, as we have extensively studied genetic and hormonal variability in our own species. Of course, neatly classifying people into two categories is difficult when social factors come into play, so we must first consider the categories 'male' and ''female'. In the first chapter, I outlined the difference between sex and gender in an attempt to separate biological and sociological classifications of people. But humans, and presumably other animals, exhibit substantial sexual differences in various biological characteristics, including those related to genetics and hormones. For example, individuals with Swyer Syndrome have an X and a Y chromosome but often develop female reproductive organs. Although these individuals are typically raised as girls, their ovaries are nonfunctional, and they generally require hormone therapy to promote the development of female characteristics. Alternatively, certain individuals with an XX genetic makeup have male physical features; this is referred to as the XX male syndrome, and all individuals expressing this genotype are also infertile. Individuals might have one or three sex chromosomes (Turner syndrome, 1 X; Klinefelter syndrome, XXY; and triple X, XXX). Intersex conditions are also possible in which the individual has characteristics of both sexes, often with both ovarian and testicular tissue.

This condition is observed in about 1 of 100,000 live births in humans. Other animals may also have intersex anatomy (Figure 4.3), although at an unknown frequency. Nearly all individuals who exhibit variation in the number of sex chromosomes are infertile and together make up less than 0.4% of the population. Their low numbers and tendency to infertility mean that such traits are unlikely to be evolutionary strategies designed to maximize fitness. Instead, such conditions are more probably developmental anomalies and cannot define individuals as a particular biological sex.

Is such genetic variability to be included in discussions about the evolution of sex and male/female strategies? Because these conditions are so infrequent in the population and because it's difficult to explain the evolution of an essentially nonreproductive sexual system, the answer is no. The duplication of chromosomes and the development of the male and female reproductive systems are not perfect, and it is somewhat surprising that it works so well most of the time. Arrangements outside the realm of typical XX females and XY males simply underscore the variability of our own physiology. It's unclear whether individuals of other animals have similar characteristics, although there's no reason to suppose they don't. However, they are unlikely to contribute such traits to the next generation because they would be generally infertile. Regardless, biological males and females are not identified based on their genetic makeup or how much

Figure 4.3 *An intersex mallard duck at Humber Bay Park, Toronto, Ontario. This female displays many male anatomical characteristics, possibly related to reduced estrogen.*
Credit: Beth Baisch/Dreamstime Photo ID 139398801.

estrogen or testosterone they have; these characteristics are simply too variable within and between species to be useful. Their sex organs are vital as they determine whether they can potentially make eggs or sperm. Consequently, this can be our only biological definition of the sexes: a biological male produces many small, mobile gametes, while a biological female produces fewer, larger, stationary gametes.

4.3 Why do gametes become so different?

The evolution of sexual strategies begins with the evolution of anisogamy or different-sized sex gametes. The investment in these cells that contribute to the genetic makeup of the following generation is the first unambiguous difference we see between males and females and sets the stage for a wide range of dissimilarities to follow. As I hope to show, parental investment, or the investment by parents into their offspring, dramatically influences the reproductive strategy of each sex. However, the evolutionary change from two morphologically similar gametes to gametes of different sizes and mobility happened over a billion years ago and isn't apparent in the fossil record. Hence, current hypotheses typically rely on evidence from mathematical models or are inferred from extant species.

Three questions need to be addressed. First, why do the gametes of two individuals have to be different in order to pair up or engage in syngamy (fertilization)? Even in those sexually reproducing organisms with no size difference in their sex cells, thus having neither males nor females, the gametes are typically divided into two (and sometimes more) types, with cells of a similar kind unable to fuse. Second, even if there is a good evolutionary reason for them to be different, why did size and mobility differences develop? In other words, during the slow process of evolutionary change, gametes go from being isogamous (similar in size but of different types) to anisogamous (different sizes). Indeed, most sexually reproducing organisms are oogamous, meaning that differences between them are magnified, with one sex responsible for producing large, stationary eggs, and the other manufacturing small, mobile sperm (or pollen if it happens to be a plant). And third, why are there commonly only two biological sexes? If there is an advantage in pairing with a different gamete, two types would be the least efficient as it would eliminate half the population from being a potential breeding partner. On the other hand, if there were 10 different types, a gamete could potentially combine with nine other kinds of partners. These are the kinds of questions that keep evolutionary biologists awake at night!

4.3.1 Why different types of gametes?

The first question is why there would be more than one type of gamete. Both gametes generally contain a single set of chromosomes (haploid) but fuse to form a zygote with two sets (diploid). Whether the mating system involves gametes of the same or different size, there usually are two distinct types. An individual automatically divides their choice for a potential mating partner by two if gametes of the same type cannot engage in fertilization. Although there are several explanations,[2] I'll summarize the two most generally accepted hypotheses. It's important to remember, though, that the rise of the eukaryotes, and thus sexual reproduction, occurred over a billion years ago, and current models are always subject to modification or even falling out of favour.

First, an individual (or gamete) may avoid combining with its own kind during fertilization to prevent inbreeding depression.[3] Inbreeding depression occurs when closely related individuals with very similar genetic makeup mate. This event increases the homozygosity of genes. In other words, individuals are more likely to get two recessive alleles that reduce the general fitness of the organism. The same reasoning can be applied to us and is why we shouldn't mate with close relatives. In general, organisms that live nearer each other are more likely to be closely related, so it's best to mate with a partner from further away. To avoid close relatives, however, you need to be able to recognize them; thus, different mating types arose.

The second reason involves a potential conflict that could occur in the zygote. In the earliest eukaryotes, the fusion of different gametes probably involved the genetic material of the two nuclei and the cytoplasmic organelles (Figure 4.4). Mitochondria, for example, must occur in the initial zygote to provide enough energy for development. However, a conflict would arise between cytoplasmic genetic elements arising from different cells; any within-cell conflict between these cytoplasmic elements would reduce

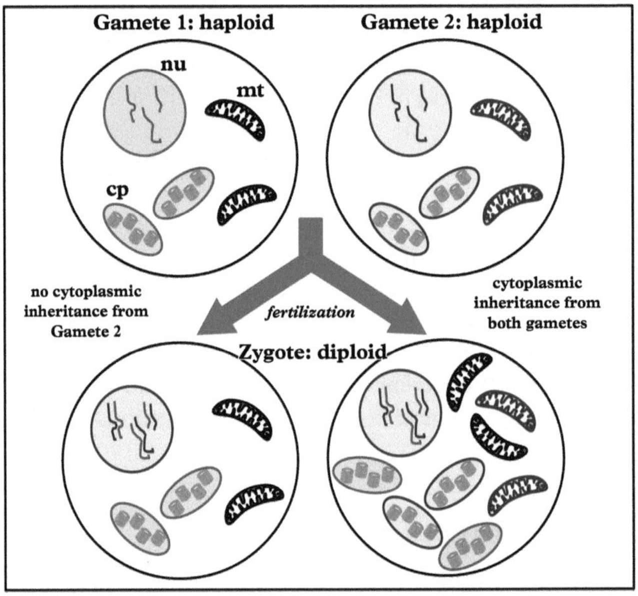

Figure 4.4 *In most sexually reproducing species, haploid gametes combine to give rise to a diploid zygote. Each parent passes on half of their nuclear (nu) genetic material. However, the mitochondria (mt) and chloroplasts (cp; in photosynthetic organisms) also have genetic material. When only one passes on the cytoplasmic organelles to the zygote, they reduce any genetic conflicts that can arise.*

the cell's fitness. To avoid such conflict, the 'organelle inheritance' model[2] suggests that it would be beneficial for nuclear DNA to force the inheritance of cytoplasmic organelles from one of the two gametes. Modelling indicates that two types of zygotes would arise if this were the case.

4.3.2 Isogamy to anisogamy

Differences in gametes that pair up may be valuable for several reasons, two of which are provided above. Many species today exist in which gamete size and mobility do not differ between individuals. Such a condition is called isogamy, the rule in unicellular eukaryotes.[4] Understanding the selection pressures that resulted in gametes going from isogamy to anisogamy, where there are two distinct sizes of gametes, is necessary to understand better why such differences arose. Let's first look at a couple of isogamous species.

Chlamydomonas is a genus consisting of nearly 300 species of single-celled green algae, generally found in freshwater or damp soil. Like many single-celled eukaryotes, many species of *Chlamydomonas* breed asexually, forming many spores. However, most also reproduce sexually, exchanging genetic information with other individuals (Figure 4.5). In the case of *Chlamydomonas reinhardtii*, the most commonly studied species, diploid parents (having two sets of chromosomes) undergo meiosis to produce 8–32 flagellated gametes having only one set of chromosomes. Two of these come together, flagellar poles touching, and fertilization occurs. Although the gametes of most *Chlamydomonas* species are morphologically similar, they are not physiologically

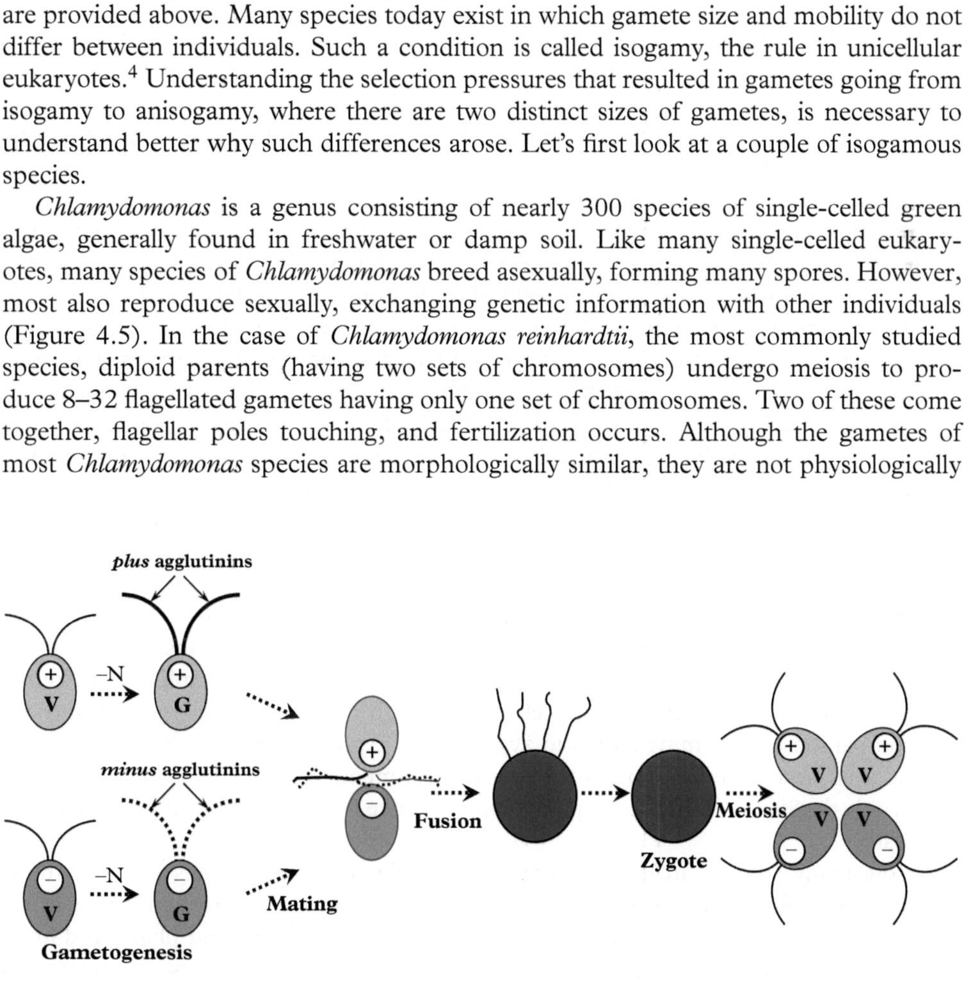

Figure 4.5 *Chlamydomonas is a good example of an organism with isogametes. This green algae's vegetative cells (V) give rise to '+' or '–' gametes (G). These two types recognize each other by the presence or absence of agglutinins on their flagellae. After fusion, zygotes undergo meiosis to produce four haploid vegetive cells.*

Credit: Hiroyuki Sekimoto; permission from Springer Nature.

the same and are generally divided into '-' and '+' types. These can recognize each other by chemicals associated with the flagella. In a few species, sexual reproduction may involve anisogamous gametes. In other words, some individuals produce a few large gametes (macrogametes or female gametes), while others make numerous smaller gametes (microgametes or male gametes). Some species take this even further. Their sex cells are unequal in size, and the female gamete withdraws the flagellum and acts as a stationary egg. You can see why this would be an excellent taxonomic group to examine when studying the evolution of eggs and sperm.

Sleeping sickness can be a serious health problem in some areas of Africa. It is caused by a single-celled parasite called a trypanosome, carried between people by tsetse flies (*Glossina spp.*). People contracting the disease often have no symptoms for months or even years after being infected, and the illness without treatment can be fatal. Indeed, sleeping sickness rivals AIDS as a cause of mortality in certain regions where it is common. After being bitten by a tsetse fly containing the trypanosome, infected individuals can develop a fever and headaches and ultimately change their behaviour; they become increasingly confused, suffer poor coordination, and finally, die. It was initially thought that trypanosomes reproduced only asexually by binary fission. However, it is now known that meiosis occurs in specific cells of trypanosomes before they come together to form a new individual. Meiosis, of course, implies sexual reproduction. There appear to be no morphological differences between haploid cells formed as a result of meiosis from different individuals; thus, they are better referred to as an isogamous sexual species.

As with sexual reproduction using anisogamous gametes, we find substantial variability in reproducing with isogametes. *Spirogyra*, introduced earlier, refers to any member of about 400 species of filamentous green algae that are commonly studied organisms in Biology classes. Being filamentous, an individual consists of a long strand of cells and usually breeds asexually by fragmentation; cells break off the parent filament, and these produce more cells by mitosis. Like the previous organism, however, it can also reproduce sexually. Sexual reproduction occurs through the joining of two cells of adjacent filaments, although it bears little resemblance to what we think of as sex. Neighbouring filaments align, and a conjugation tube forms between two adjacent haploid cells, known as isogametes. The contents of one cell pass through the tube and merge with those of its companion cell, forming a diploid zygote. This zygote develops a thick wall around it and then overwinters, waiting for better environmental conditions to emerge. In this case, only the zygote is diploid; haploid filaments arise from the zygote after it undergoes meiosis.

How many species are isogamous? Most single-celled eukaryotes are isogamous, while most multicellular eukaryotes are anisogamous, although many exceptions exist.[4] For example, there are isogamous species in a few multicellular land plants and green algae, while certain unicellular green algae are anisogamous. Of course, it's somewhat challenging to determine the types of gametes found in many organisms as the reproductive mode of a large number, especially unicellular organisms, is still unknown. However, we know that anisogamy is the general condition in all known animals.

The variation in gamete morphology and behaviour in groups of closely related organisms, such as green algae[5] or fungi,[6] may hint at why certain gametes may be favoured under specific environmental conditions. However, most of our understanding of the transition from isogametes to anisogametes comes from mathematical models. The best known of these was developed in the early 1970s by Geoff Parker and his coworkers,[7] and various modifications by Parker and others have resulted in a generally accepted hypothesis.[8] Named the PBS model (after its original authors), it is based on competition between gametes to find other gametes with which to undergo syngamy. The model assumes that individuals have a certain amount of energy to put into gamete production. Therefore, they can make a few large, many smaller, or a moderate number of intermediate-sized gametes. Fusion between gametes is random and not size-related, and the fitness of the zygote produced through the fusion between two gametes increases with size. Given these conditions, the simulation results in two sizes of gametes—very small and very large. This outcome is termed an **evolutionarily stable strategy** such that it can't be bettered (in terms of fitness) by any other strategy. Two other developments serve to reinforce size differences between gametes. First, while the larger gamete invests in resources necessary in early development, the smaller gamete is more capable of movement.[9] In addition, the process speeds up if the larger cells release a pheromone that facilitates their discovery, and the smaller cells develop methods to detect that pheromone.[10]

Perhaps an analogy will help better illustrate this. Picture yourself on a nice sunny day, walking in a forest. You stroll for hours, taking in the sights and smells around you, lost in your thoughts. But as you return to the moment, you realize you don't know where you are or how to get out of the forest—you're lost. Don't panic! You told your partner that you were going for a walk in this particular forest, and surely, they would eventually come for you. So, the first thing you should do is stay put. Walking further may make you even more lost, and it's easier for somebody to find you if you remain in one place. How can you increase your chance of being found? Put on something bright so you can be spotted more easily. Or yell. Or better yet, turn on the locator function on your phone. These will all make you more noticeable and work at different distances. And how can your partner increase the chance of finding you? It is best to invite all their friends to search as well. Having more eyes and covering a larger area will aid in their search. Also, ensure they don't forget to bring their own device to locate your phone.

The evolution of different gamete sizes is a classic case of disruptive selection, one of three selection processes described in nearly all introductory readings of natural selection (the other two being stabilizing and directional selection; Figure 4.6). Here's how it works. The population of gametes starts with all individuals approximately the same size. There is some variation—a few gametes are smaller, some are larger, and most are average. This describes the normal distribution for most characteristics. Think of the height of individuals in a population—most would be of medium height, although there would be some very tall and very short individuals. In disruptive selection, any advantage of being small or large will increase a gamete's chance of survival at the expense of the ones of medium size. Gametes of intermediate size cost more to produce than the smaller ones but don't contain the resources that the larger ones do. Thus, they are at a disadvantage and decrease in numbers. After a while (in evolutionary time, this

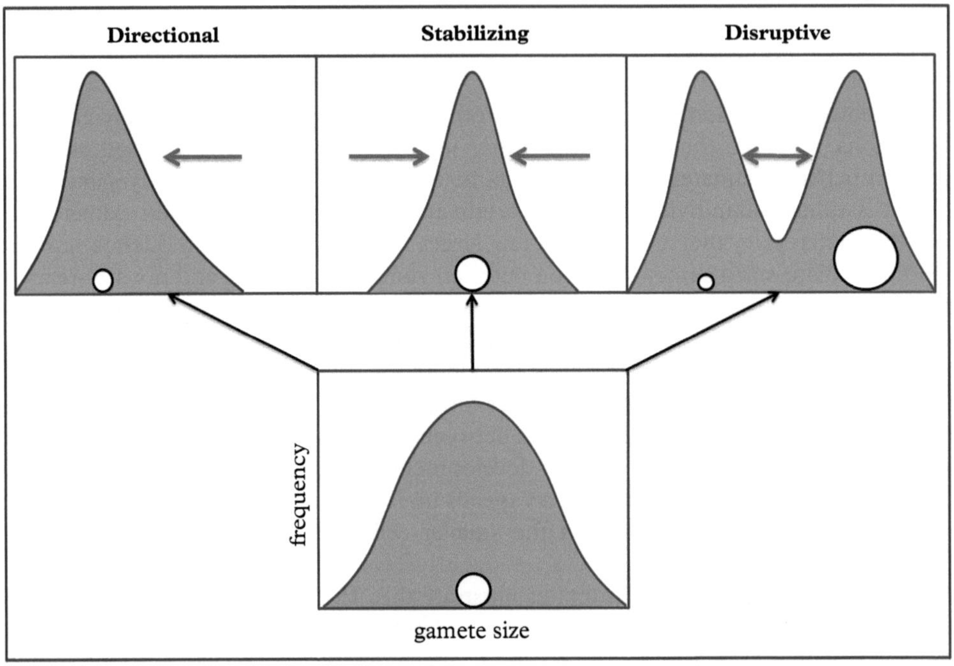

Figure 4.6 *Three types of natural selection. In stabilizing selection, individuals with average characteristics have an advantage. In directional selection, individuals having the feature at one extreme are elected. In disruptive selection, the process by which gametes are thought to have evolved, individuals having the characteristic at either extreme are selected.*

could be hundreds of thousands of years), only two types of gametes are expected to remain—large, stationary ones and small, mobile ones.

4.3.3 Are two sexes the rule?

When two gametes are of equal size or isogamous, they are usually referred to as **types**. When they are of different sizes or anisogamous, they are typically called separate **sexes**, the male being the smaller gamete and the female being the larger. So, the evolution of sex-specific sexual strategies really begins at this point since, for the first time, there are clear and predictable differences between males and females. But it's fair to ask whether there are ever more than two types or sexes in sexually reproducing organisms.

Slime moulds (*Dictyostelium discoideum*), or social amoebae, separated early from the animal lineage, even before fungi split off. Slime moulds are relatively unique for several reasons. First, they can shift between unicellularity and multicellularity during the course of their lives. They can also reproduce asexually, either by fission of singular cells or when food is scarce, through differentiation into fruiting bodies that produce spores, or sexually, fusing with other cells and forming a diploid zygote. The odd thing about them is that there are three mating types—I, II, and III.[11] Each type can fuse with either of the other two types but not its own. Thus, zygotes can be type I–II, type I–III, or type II–III. After the fusion of two kinds of *D. discoideum*, the resulting diploid zygote releases a chemical that attracts other cells. After assisting with producing a hard outer shell around the zygote, these cells are promptly ingested. The microcyst, thus formed, undergoes meiosis and multiple mitotic events to generate a large number of haploid individuals.

Many of the ciliates often have more than two mating types. The cells do not fuse but rather exchange nuclei through a conjugation tube. Ciliates have two nuclei—a micro- and macronucleus. Only the micronucleus undergoes meiosis and is exchanged; thus, it is the 'sexual' nucleus. For example, *Tetrahymena thermophila*, a free-living ciliate protist, has seven different mating types and can reproduce asexually and sexually. Sexual reproduction comes when two cells find themselves in food-limited conditions.[12] In this case, fusion can occur between any of the different types, effectively increasing your chance of finding a mate to 6/7 instead of 1/2. Oddly, the mating types of the parents have no relationship to the mating type of the offspring; each organism contains the DNA of all seven forms. The mating type is determined as parts of DNA are shut off. Although *Tetrahymena* bypasses the production of 'male' and 'female' gametes, it produces stationary and migratory pronuclei that are exchanged between complementary mating types.

Some fungal species can have thousands of mating types and breed with just about any individual they bump into.[13] Although I deal primarily with eukaryotic animals, fungi are an appropriate group to study when examining the evolution of sexual reproduction, as they have a range of reproductive methods. These include asexual and sexual modes of reproducing, isogamous or anisogamous gametes, and a variable number of mating types from one to thousands.

What about animals? In 2016, an article was published with the intriguing title 'The sparrow with four sexes'.[14] White-throated sparrows (*Zonotrichia albicollis*) have two plumage morphs—one with white stripes on the tops of their heads and the other with tan stripes (Figure 4.7). The two morphs differ not only in their appearance but also behaviourally. Singing ability, parental devotion, and even aggressiveness are all different. These differences can be traced to a chromosome that has undergone an inversion where part of chromosome 2 flipped over on itself.[15] Instead of the genes on the chromosome reading A B **'C D E'** F G, they read A B **'E D C'** F G. But this simple example minimizes the change as over a thousand genes were involved. Crossing over couldn't happen between these genes because they couldn't line up across from each other during meiosis; therefore, they constituted a 'superfamily of genes' and were all inherited together. This arrangement would explain the anatomy and behavioural differences between the two morphs. So, do they have four sexes? Unusually, the two morphs mate assortatively—white-striped males only mate with tan-striped females, and vice-versa. Additionally, as it turns out, the behavioural differences associated with the chromosome inversion explain why opposite morphs attract—they complement one another. For example, birds with white stripes are better at getting mates but make lousy parents, while those with tan stripes are the opposite. So, males and females are more successful reproductively when they mate with a different morph. However, these aren't different sexes but rather different mating types—there are two types of males and two types of females.

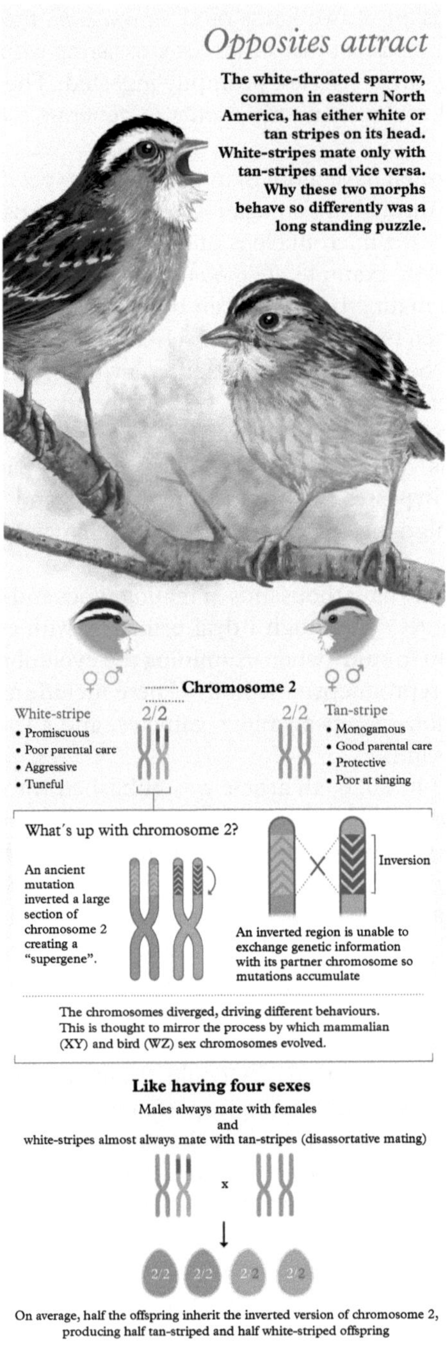

Figure 4.7 *White-throated sparrows have two morphs formed because of an inversion on chromosome 2. The two morphs are anatomically and behaviourally different and nearly always mate with individuals of the opposite morph. This preference limits the potential pool of mating partners to 25% of the population. However, there are still only two sexes, male and female; in this case, each has two mating types.*

Credit: Carrie Arnold; permission from Springer Nature.

4.3.4 Are two individuals the rule?

Throughout this discussion, I've implied that males and females, defined by their gametes, occur in separate bodies. But do different gametes necessarily imply different individuals? In most cases, eggs and sperm are produced by two organisms. However, in many cases, the machinery required to produce gametes of both sexes occurs concurrently or consecutively in one body. Hermaphrodites are individuals that can have both male and female reproductive organs and, therefore, can make eggs and sperm during their lifetimes. Hermaphroditism is not uncommon and occurs in a diverse array of phyla, suggesting that it has evolved repeatedly. In animals, hermaphroditism is found in only 5%–6% of all species but is much more common in plants, found in almost 94% of species.[16] I will discuss hermaphrodites in more detail later; still, I think it's necessary to say something now as they seem to contradict the 'two sexes having two different strategies' approach to sexual reproduction. If both sexes are contained within a single individual, how can one speak of male and female ways of promoting their genes?

There are certainly advantages to switching sexes over the course of your life, and I discuss a few species in which individuals can do this in Chapter 7. Individuals that are sequential hermaphrodites (i.e. they change sex at some point in their lives) face many of the same problems in reproduction as species that always have separate sexes. The search costs for potential mates would likely decrease, but the metabolic machinery required to produce male and female gametes over their lifetime would undoubtedly be higher. Simultaneous hermaphroditism, when an individual can manufacture male and female gametes at the same time, can face other advantages and disadvantages. Such species are divided into two groups: those that can self-fertilize and those that can't. However, the ability to self-fertilize (selfing) is rare among simultaneous hermaphrodites. If offspring are of low quality when you mate with close relatives, consider the costs of producing offspring by mating with yourself. The main advantage of self-fertilization was proposed by Darwin,[17] who suggested that producing poor-quality offspring may be better than not breeding at all (the reproductive assurance hypothesis).

Regardless, the driving force for hermaphroditism is a species' mode of locomotion and its encounter rate with other individuals of its species.[18] Stationary individuals (like plants) are much more likely to be hermaphroditic because encounter rates with other individuals of the same species are low. While this is especially true for organisms that can self-fertilize, it also makes things easier for those who cannot, enabling them to combine gametes with any other individual. The degree to which inbreeding depression is detrimental to offspring is considered one of the most important factors favouring the evolution of the sexes in different, rather than the same, individuals.[18]

4.4 Coming together

The coming together of haploid gametes is the first interaction between two sexually reproducing individuals in terms of maximizing their reproductive fitness. Suppose it happens to be between anisogamous gametes (i.e. ones that can be classified as carrying male and female genotypes). In that case, fertilization can be one of the first instances of conflict or co-operation between the sexes. Co-operation may occur because both males and females 'want' fertilization to happen; that is why they produce gametes in the first place. Conflict, however, might occur because the female can often influence the success of the male with whom she mates. Obviously, she doesn't want any old gamete. Since she's invested so much energy into the egg, she wants to combine her genetic material with a high-quality individual. However, before addressing the co-operation/conflict strategies of males and females, you should understand how the gametes come together to understand better how females may influence their mate choice. But remember, as I said in the introduction, I am not a molecular biologist, so I'm limiting this discussion to a few well-studied examples. For those wanting to know more about the molecular processes involved in fertilization, please consult any articles that review this topic.[19,20]

The larger gamete releases chemicals or pheromones in many organisms, making it easier for the smaller, mobile gamete to find it. I'll focus on the algae, as gamete attraction has been extensively studied in this group. Different species of algae can have isogamous, anisogamous, or even oogamous gametes, and it's known that pheromones play an important role in gamete attraction.[10] While the isogamous gametes of *Chlamydomonas reinhardtii* seem to come together randomly, the MT- gamete of the closely related *Chlamydomonas allensworthii* releases a lurlenic acid derivative that MT+ gametes can detect. *Oedogonium* is a freshwater, filamentous genus of green algae that produces anisogamous gametes, and the eggs also attract sperm by releasing a pheromone. Lastly, the brown algae *Ectocarpus siliculosus* might represent an intermediate evolutionary stage in which female gametes are larger than those of males, but both gametes are mobile. However, only the eggs of this species release a pheromone to attract sperm. Many other examples of pheromone attraction exist in various algal species, but you get the picture.

Pheromone attraction of gametes is also essential in other eukaryotes such as ciliates,[21] fungi,[22] and mammals.[23] However, finding each other is just the first stage. After being attracted to one another, the two gametes must fuse to initiate fertilization and subsequent development. While such processes are complex and differ in distinct groups, they all begin when gametes make contact. In anisogamous gametes, fusion requires the appropriate membrane proteins on the sperm and corresponding receptors on eggs. After fusion occurs between the gametes, the genetic material of the two organisms must be combined. Both processes are crucial if fertilization is to be successful.

It is essential to clarify the usage of specific terms, as they are often used specifically for the organism being examined. 'Plasmogamy' refers to the fusion of membranes of two haploid cells and the subsequent combining of material from the two cells. After this occurs, the resulting zygote has two sets of genes, each in its own pronucleus. Before becoming functionally diploid, those pronuclei must merge into a common nucleus in a process called **karyogamy**. **Syngamy** incorporates both plasmogamy and karyogamy; when it occurs between two unrelated cells, syngamy is often used interchangeable with fertilization (Figure 4.8).

Fusion of the membranes of two cells, and later of two nuclei, doesn't occur in prokaryotes. Remember, prokaryotes were (and are) asexual and have little need for fusion with neighbouring prokaryotes. They use fimbriae (short processes) or pili (long processes) that extend from their cell membranes for locomotion, sticking to other objects, and sometimes even in DNA transfer between cells. Membrane fusion occurs

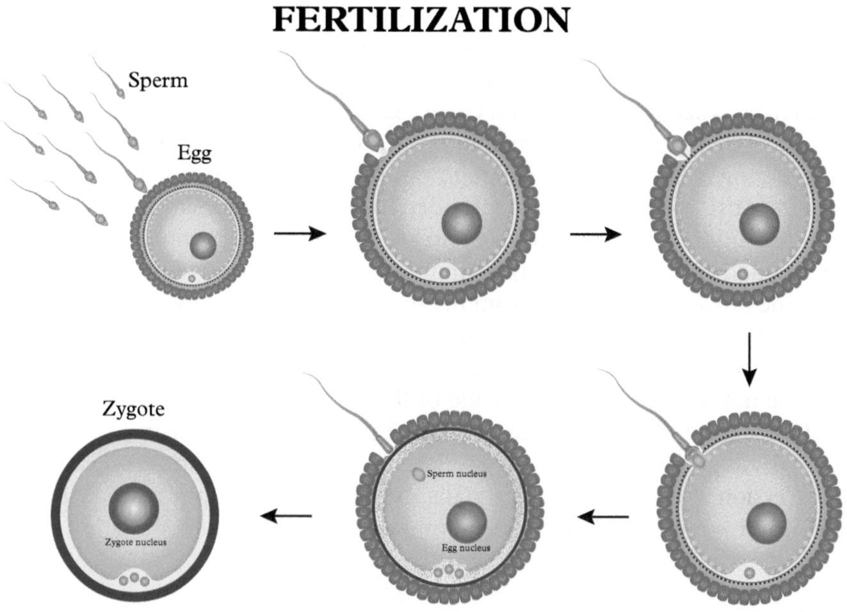

Figure 4.8 *Male and female gametes come together in a process known as fertilization or syngamy. First, the membranes of the two cells must recognize and fuse, a process that often relies on chemical attractants released by the larger cell and recognition proteins on the surface of each cell; this is called 'plasmogamy'. Once the nucleus of the smaller cell is released into the cytoplasm, it uses spindle fibres to move to the female nucleus. More recognition proteins enable the two pronuclei to fuse in a process known as 'karyogamy'.*

Credit: Veerathada Khaipet | Dreamstime Photo ID 148692216.

in eukaryotes when gametes come together and occurs during the formation of other eukaryotic structures, such as skeletal muscle fibres. Plasmogamy is an area of active research, and the necessary membrane proteins often differ between organisms. In each case, however, proteins on the surface of one cell, or gamete, bind to receptors on the membrane of the other cell, initializing fusion events.

Gene deletion studies have been particularly instrumental in uncovering the proteins necessary for the fusion of gametes of different eukaryotes. Strains of organisms that lack a particular gene (or have one that is nonfunctional) are produced, and the missing protein is determined by comparing them with normal cells. Since the first eukaryotic **fusogen**, a protein instrumental in membrane fusion of gametes or other cells was discovered 20 years ago in the nematode *C. elegans*,[24] many other fusogens have since been identified. (**Fusexins** are a superfamily of fusogen proteins explicitly involved in gamete fusion.) Although different species may have different fusogens, the transmembrane protein HAP2/GCS1 is highly conserved and known to function in the fusion of gametes in a diverse range of eukaryotes, including protozoans, plants, cnidarians, molluscs, and arthropods, suggesting that they evolved early in eukaryotes.[25] In some cases,

HAP2/GSC1 appears to be necessary in only one of the gametes (generally the sperm), while in other species, both require it. Recently, a gene homologous to the HAP2/GSC1 fusogen was discovered in an archaean prokaryote,[26] and these homologues stimulate fusion activity when transfected into eukaryotic cells.

Once enclosed by a single cell membrane, the two nuclei are referred to as 'pronuclei', and the organism is effectively, although not functionally, diploid. Again, 'karyogamy' refers to the fusion of the two pronuclei, derived from the haploid stage of entire single-celled organisms (e.g. protozoans, some yeast, or some fungi) or haploid cells that have arisen through meiosis (gametes) in multicellular organisms. This process generally involves the movement of nuclei from the two cells, controlled by microtubules (in animals) or actin (in plants) that are organized by the centriole of the cell.[26] Although the physical processes involved in karyogamy have been extensively studied, its molecular aspects are not well-understood outside of the fungi, where it has been investigated in detail. In this group, it is clear that the KAR2 protein plays an instrumental role in the fusion of the nuclei, while a homologous protein, BiP/GRP78 (better known for its role in the progression of certain neurological diseases and cancer), appears to be homologous to KAR2 and is important in the fusion of mammalian pronuclei.[20]

4.5 Summary

Males and females of various species differ in many ways and are relatively easy to tell apart. But this isn't always the case. In many organisms, the sexes are impossible to distinguish purely by external appearance. However, males consistently produce the smaller sex cells, while females generate the larger ones. This also means that, for the same energy, males can manufacture many sex cells, while females produce far fewer. Additional characteristics, including genetics and hormones, may also differ between males and females. However, because of their variability within a species, and differences between species, none are as consistent as gamete size when defining 'males' and 'females' for biological purposes.

Why do gametes have to be different in the first place? Even if there are no size differences between sex cells, they must differ in other ways, or they won't join up. Perhaps these differences arise to prevent inbreeding. Or perhaps differences are to ensure one gamete supplies all the cell organelles. However, once differences occur, computer models suggest the ability to find each other leads to size differences; differences in mobility and pheromone attraction augment these size divergences. And while many species have more than two kinds of gametes, these are generally not regarded as different sexes but rather distinct types. It is important to note that male and female gametes aren't always found in separate individuals. Hermaphrodites can produce gametes of both sexes and are typically found in species with lower encounter rates than other individuals. However, this advantage is at least partially offset by the lower quality of offspring potentially produced because of selfing.

Finding each other is often related to chemical signals released by one of the gametes, most often the egg. Recognition proteins located in the cell membranes of both male and female gametes are instrumental in the sex cells coming together and fusing. I briefly discussed some of the more critical proteins essential in syngamy or this fertilization process.

References

1. Ford AT, LeBlanc GA. Endocrine disruption in invertebrates: A survey of research progress. Environ Sci Technol [Internet]. 2020 Oct [cited 2022 Oct 16];54(21):13365–13369. Available from: https://doi-org.proxy.library.upei.ca/10.1021/acs.est.0c04226
2. Billiard S, López-Villavicencio M, Devier B, Hood ME, Fairhead C, Giraud T. Having sex, yes, but with whom? Inferences from fungi on the evolution of anisogamy and mating types. Biol Rev Camb Philos Soc [Internet]. 2011 May [cited 2022 Oct 14];86(2):421–442. Available from: https://doi.org/10.1111/j.1469-185X.2010.00153.x
3. Charlesworth D, Charlesworth B. The evolution and breakdown of S-allele systems. Heredity 1979;43(1):41–55.
4. Lehtonen J, Kokko H, Parker GA. What do isogamous organisms teach us about sex and the two sexes? Philos Trans R Soc Lond B Biol Sci [Internet]. 2016 Oct [cited 2022 Oct 20];371(1706):1–12. Available from: https://doi.org/10.1098/rstb.2015.0532
5. Kim GH, Lee IK, Fritz L. Cell-cell recognition during fertilization in a red alga, *Antithamnion sparsum* (Ceramiaceae, Rhodophyta). Plant Cell Physiol [Internet]. 1996 Jul [cited 2022 Oct 20];37(5):621–628. Available from: https://doi.org/10.1093/oxfordjournals.pcp.a028990
6. Merlini L, Dudin O, Martin S. Mate and fuse: How yeast cells do it. Open Biol [Internet]. 2013 Mar [cited 2022 Nov 20];3(3):130008. Available from: https://doi.org/10.1098/rsob.130008
7. Parker GA, Baker RR, Smith VGF. The origin and evolution of gamete dimorphism and the male-female phenomenon. J Theor Biol 1972 Sep;36(3):529–553.
8. Lehtonen J. The legacy of Parker, Baker and Smith 1972: Gamete competition, the evolution of anisogamy, and model robustness. Cells [Internet]. 2021 Mar [cited 2022 Nov 21];10(3):573. Available from: https://doi.org/10.3390/cells10030573
9. Dusenbury DB. Selection for high gamete encounter rates explains the evolution of anisogamy using plausible assumptions about size relationships of swimming speed and duration. J Theor Biol [Internet]. 2006 Jul [cited 2022 Nov 22];241(1):33–38. Available from: https://doi.org/10.1016/j.jtbi.2005.11.006
10. Frenkel J, Vyverman W, Pohnert G. Pheromone signaling during sexual reproduction in algae. Plant J [Internet]. 2014 Aug [cited 2022 Nov 22];79(4):632–644. Available from: https://doi.org/10.1111/tpj.12496
11. Okamoto M, Yamada L, Fujisaki Y, Bloomfield G, Yoshida K, Kuwayama H, et al. Two HAP2-GCS1 homologs responsible for gamete interactions in the cellular slime mold with multiple mating types: Implication for common mechanisms of sexual reproduction shared by plants and protozoa and for male-female differentiation. Dev Biol [Internet]. 2016 Jul [cited 2022 Nov 23];415(1):6–13. Available from: https://www.ncbi.nlm.nih.gov/pmc/articles/PMC4910948/ DOI: 10.1016/j.ydbio.2016.05.018
12. Cervantes MD, Hamilton EP, Xiong J, Lawson MJ, Yuan D, Hadjithomas M, et al. Selecting one of several mating types through gene segment joining and deletion in *Tetrahymena thermophila*. PLoS Biol [Internet]. 2013 Mar [cited 2022 Nov 20];13(10):e1001518. Available from: https://doi.org/10.1371/journal.pbio.1001518

13. Casselton LA, Olesnicky NS. Molecular genetics of mating recognition in basidiomycete fungi. Microbiol Mol Biol Rev [Internet]. 1998 Mar [cited 2022 Nov 22];62(1):55–70. Available from: https://doi.org/10.1128/MMBR.62.1.55-70
14. Arnold C. The sparrow with four sexes. Nature [Internet]. 2016 Nov [cited 2022 Feb 8];539:482–484. Available from: https://doi.org/10.1038/539482a
15. Tuttle EM, Bergland AO, Korody ML, Brewer MS, Newhouse DJ, Minx P, et al. Divergence and functional degradation of a sex chromosome-like supergene. Curr Biol [Internet]. 2016 Feb [cited 2023 Feb 8];26(3):344–350. Available from: https://doi.org/10.1016/j.cub.2015.11.069
16. Jarne P, Auld JR. Animals mix it up too: The distribution of self-fertilization among hermaphroditic animals. Evolution [Internet]. 2006 May [cited 2002 Dec 2];60(9):1826–1824. Available from: https://doi.org/10.1111/j.0014-3820.2006.tb00525.x
17. Darwin CR. The effects of cross and self fertilisation in the vegetable kingdom. 2nd ed. London: John Murray. In: Wyhe J editor,. The Complete Work of Charles Darwin Online. [Internet]. 2002. Available from: http://darwin-online.org.uk/content/frameset?pageseq=1&itemID=F1251&viewtype=text
18. Eppley SM, Jesson LK. Moving to mate: The evolution of separate and combined sexes in multicellular organisms. J Evol Biol [Internet]. 2008 Mar [cited 2022 Dec 2];21(3):727–736. Available from: https://doi.org/10.1111/j.1420-9101.2008.01524.x
19. Evans JP, Sherman CD. Sexual selection and the evolution of egg-sperm interactions in broadcast-spawning invertebrates. Biol Bull [Internet]. 2013 Aug [cited 2022 Nov 22];224(3):166–183. Available from: https://doi.org/10.1086/BBLv224n3p166
20. Fatema U, Ali MF, Hu Z, Clark AJ, Kawashima T. Gamete nuclear migration in animals and plants. Front Plant Sci [Internet]. 2019 Apr [cited 2022 Nov 26];10:517. Available from: https://doi.org/10.3389/fpls.2019.00517
21. Luporini P, Alimenti C, Ortenzi C, Vallesi A. Ciliate mating types and their specific protein pheromones. Acta Protozool 2005;44(2):89–101.
22. Bölker M, Kahmann R. Sexual pheromones and mating responses in fungi. Plant Cell [Internet]. 1993 Oct [cited 2022 Nov 6];5(10):1461–1469. Available from: https://doi.org/10.1105/tpc.5.10.1461
23. Eisenbach M, Giojalas LC. Sperm guidance in mammals—An unpaved road to the egg. Nat Rev Mol Cell Biol [Internet]. 2006 Apr [cited 2022 Nov 15];7(4):276–285. Available from: https://europepmc.org/article/med/16607290 DOI: 10.1038/nrm1893
24. Mohler WA, Shemer G, del Campo JJ, Valansi C, Opoku-Serebuoh E, Scranton V, et al. The type I membrane protein EFF-1 is essential for developmental cell fusion. Dev Cell [Internet]. 2002 Mar [cited 2022 Nov 26];2(3):355–362. Available from: https://doi.org/10.1016
25. Vance TD, Lee JE. Virus and eukaryote fusogen superfamilies. Curr Biol [Internet]. 2020 Jul [cited 2022 Nov 26];30(13):R750–R754. Available from: https://doi.org/10.1016/j.cub.2020.05.029
26. Moi D, Nishio S, Li X, Valansi C, Langleib M, Brukman NG, et al. Discovery of archaeal fusexins homologous to eukaryotic HAP2/GCS1 gamete fusion proteins. Nat Commun [Internet]. 2022 Aug [cited 2022 Nov 6];13:3880. Available from: https://doi.org/10.1038/s41467-022-31564-1

5

Sexual Reproduction Is Costly

5.1 Sex isn't always needed

Indonesia is made up of over 14,000 islands. On a few of these, you might be lucky enough to see a Komodo dragon (*Varanus komodoensis*), also known as a Komodo monitor (Figure 5.1). For the most part, Komodo dragons are typical vertebrates. Males fight each other for access to females, and the female usually mates with the strongest available male. Besides the dangers faced while fighting other males, copulation itself can be quite dangerous to the male because the female may be pretty aggressive, especially during the initial stages of courtship. The male has an intromittent organ with two parts called hemipenes (each is referred to as a hemipenis) that are functionally similar to the mammalian penis. If all goes well, he will insert one hemipenis into the female, transfer his sperm, and hopefully fertilize her eggs. Unusually for reptiles, the male may then hang around so that he mates exclusively with that female, thus siring her entire brood. Such a process requires a substantial commitment in terms of time and energy, not only in fighting and mating, but also in the anatomical features that make an individual more successful at both.

Female Komodo dragons can also produce offspring without any contribution from males.[1] Two female komodos, held in zoos without access to males, laid clutches of eggs, many of which later hatched successfully. At first, keepers thought the females had stored sperm from an earlier encounter with males. However, genetic testing demonstrated that there was no contribution by any individual other than the mother.

So, females of this species can reproduce without all the time and trouble involved in breeding and produce fully viable offspring with all their mother's genes. As we saw earlier, female whiptail lizards have gone one step further; they have eliminated males altogether, and the entire population consists only of females. However, this is unusual, especially in vertebrates. But why? If individuals of certain species can reproduce without having sex, why have sex at all? Why not avoid all the costs and produce offspring that have 100% of your genes? There are at least three significant costs to sexual reproduction—the costs of meiosis, males, and mating. Let's look at these in more detail.

The Evolution of Sex. Kevin Lee Teather, Oxford University Press. © Kevin Lee Teather (2024).
DOI: 10.1093/9780191994418.003.0005

Figure 5.1 *The Komodo dragon in its natural environment on Rica Island, Indonesia. In zoos, females have been known to give birth parthenogenetically—no male required! Adult females produce clones of themselves.*
Credit: Sergey Uryadnikov/Dreamstime Photo ID 101736607.

5.2 The cost of meiosis

We've looked at meiosis previously, so you should know what it does and, at least potentially, how it evolved. Remember that meiosis is a process found in all sexual reproducers and, thus, is a consistent characteristic of sex. To quickly review, there are two significant processes in meiosis, at least as far as we are concerned. First, it reduces the complement of chromosomes in a cell, deriving four haploid cells from one that is diploid. One of these haploid cells must interact with that of another individual (or rarely the same individual) to produce offspring. Meiosis effectively dilutes the genes passed on by the reproductive individual by 50%. The second process is the crossing over of genetic material from the chromosomes of both parents and the subsequent random allocation of chromosomes to each gamete. In other words, each haploid cell ends up with a unique genotype because of the recombination of alleles. The extent of mixing is variable and depends on the degree to which genes recombine. The genetic makeup of progeny is further diversified when the gametes of two individuals mix.

Meiosis was almost certainly present in the earliest eukaryotes[2] and likely evolved from transformation in the prokaryotes. Transformation permits one bacterium to take in DNA floating in the extracellular environment, having come from another closely related bacterium. The DNA taken in this way probably functioned as a template to repair damage or deleterious mutations on the circular chromosome of the competent bacterium. It had the added benefit, probably longer-term, of increasing the variability of the genotype, either by introducing new alleles into the homologous region of DNA or even by introducing new genes where the recombination wasn't in a homologous area. In modern meiosis, allele exchange only occurs in homologous regions.

George Williams, one of the leading evolutionary biologists of the last 100 years, is partially responsible for current evolutionary ideas regarding senescence, menopause, levels of selection, and medicine. He also recognized that when compared with asexual reproduction, sexual reproduction was costly.[3] Williams suggested that the main cost of sex was associated with meiosis, as an organism had to combine its genetic material with another individual to produce young. In other words, only half their genotype was represented in their gamete, which came together, at fertilization, with the gamete of their mating partner to produce a zygote. This process has also been termed **genome dilution**. While asexual individuals pass on their entire genomes through mitosis, sexual organisms could contribute only 50% of theirs because of meiosis.

It was later recognized that, from an evolutionary viewpoint, the only reduction that would matter would be in the genes contributing to the reproductive mode[4]. An example might clarify this.[5] Consider a sexual species in which sex is controlled by the recessive gene 'a' so that all individuals have an 'aa' genotype. Now assume that a dominant mutation 'A' arises that permits females to reproduce parthenogenetically or asexually. In the population, there are now 'Aa' asexual and 'aa' sexual individuals. There can be no AA individuals since Aa females only produce identical copies of themselves. Because the two lineages are reproductively, and thus genetically, isolated from each other, an asexual mutant has no way of increasing in the population of sexual reproducers. For this reason, the original genome is never diluted. This result is also the case (although it takes a little longer) if a dominant, rather than recessive, gene codes for sexual reproduction.[5] This outcome doesn't mean there's no cost to meiosis, but simply that this cost has little to do with genome dilution. It also doesn't negate the twofold cost of males, discussed in Section 5.3.

In addition to dividing the number of chromosomes in half and reducing an individual's genetic input into their progeny, meiosis can also split up good sets of genes that might benefit the offspring. This is the actual cost of meiosis. However, we need to look at meiosis more closely to understand how this can happen. Recall synapsis and the subsequent events that occur during Prophase I: paired chromosomes align, and parts of them can break and switch strands or 'cross over'. The newly formed chromosome now has a unique mixture of alleles derived from each parent. Because of this recombination of alleles, all cells produced through meiosis are different. Thus, meiosis results in all offspring being different from each other and either of their parents. And this can be good or bad, depending on how you look at it.

If an organism can survive and reproduce under certain environmental conditions, it must be well adapted to those conditions. In other words, their combination of genes gives them the characteristics they need to do well in that habitat. Let's imagine that part of this success is due to the alleles of two genes located very close to each other on one of the chromosomes and working well together in the adult. Usually, these alleles pass through meiosis together because of their proximity to each other, and both end up together in the egg or sperm. However, there is a chance that the regions of DNA that code for these genes may get split up during recombination, so the gamete ends up with a less effective combination of those genes. Therefore, the main problem that may result from meiosis is breaking up what might be advantageous genetic arrangements. Otto[6] likened this to reshuffling a good poker hand. If you had three aces in your hand for one game and were allowed to keep them or throw them in and get new ones for the next match, what would you do?

Perhaps a hypothetical example will help make this more straightforward. Let's say that females of a particular species find males with blue eyes and big ears desirable and, thus, are more likely to breed with them. The gene for eye colour has two alleles—one that gives you blue eyes (which are good to have, if you are a male at least) and another that gives you green eyes. The gene for ear size also has two alleles—big ears (good) and small ears (not so good). It just happens that these genes sit right next to each other on the chromosome. It would be great for males if the genes for blue eyes and prominent ears alleles were always inherited together, but meiosis reduces this probability. Generally, these alleles pass through meiosis together, and the male offspring receive both traits. However, every once in a while, these genes are split up during recombination, and a male offspring may have blue eyes and small ears, brown eyes and big ears, or—God forbid—brown eyes and small ears. In other words, they end up with combinations of characters that make them less appealing to females. Even though it would be better to keep those characters together, crossing over during meiosis could potentially separate them. And therein lies the actual cost of meiosis.

5.3 The cost of producing males

It's easy to confuse the 'cost of males' with the 'cost of meiosis'; previous studies have not helped matters, often messing them up or even using them interchangeably. This confusion stems from the implications of sexual reproduction laid out by George Williams, who pioneered the 'genome dilution' argument and John Maynard Smith, who outlined the 'cost of males' scenario.[7] One of the problems is that both hypotheses proposed a twofold advantage for individuals that reproduce asexually, and thus, they should quickly take over populations should they arise. As outlined above, this doesn't work when talking about genome dilution, but what about the 'cost of males'? Let's look at this more closely.

An organism uses the energy it takes in for a variety of processes. There is the growth and maintenance of the body and body structures. There are different activities, such as feeding and the avoidance of predators. And, of course, there is the energy devoted to reproduction. Now imagine a fish with enough energy to produce two offspring after dealing with all the other activities, then dies. A female has done her evolutionary job by producing just two offspring that survive long enough to breed themselves; generating two progeny results in a stable population—one offspring to replace herself and one to replace her mate. If the population has the same number of males and females, mothers will produce, on average, one daughter and one son over their lifetimes.

Now imagine a mutation that permits a female fish of the same species to breed asexually. In other words, she doesn't need a male to fertilize her eggs. She produces two daughters using the same amount of energy for reproduction (plus avoiding the danger, time, and costs of finding and copulating with a suitable mate). In other words, both females (the mutant and nonmutant) produce the same number of offspring, using the same amount of energy. However, in the following generation, the two daughters of the mutant asexual fish have four offspring (all daughters), while the daughter of the sexual fish produces only two offspring, and the male, of course, doesn't produce any (Figure 5.2a). Given that the asexual mutant is the same in other respects (i.e. it competes and uses resources as effectively as sexual individuals), the difference in offspring number means that the descendants of the asexual one will soon wipe out descendants of the original sexual fish. Before you think this is an entirely hypothetical situation, it (or something similar) can happen. Redbelly dace (*Chrosomus eos*) and finescale dace (*Chrosomus neogaeus*), two closely related species, can mate with each other and produce asexual offspring capable of having offspring themselves.[8] However, the hybrid doesn't take over, possibly because it may be a poorer swimmer than either parent species. In other words, it may be physiologically inferior, wiping out any breeding advantage it might have.

How long will it take for the asexual species to take over? If the carrying capacity of our fish population size were a hundred thousand, and there were no differences in the survival between the offspring of asexual and sexual fish, there would be fewer than two sexual individuals after 33 generations (Figure 5.2b). Fish generations are often not very long, and it might be easy to see such a change in our lifetime. Given that we usually

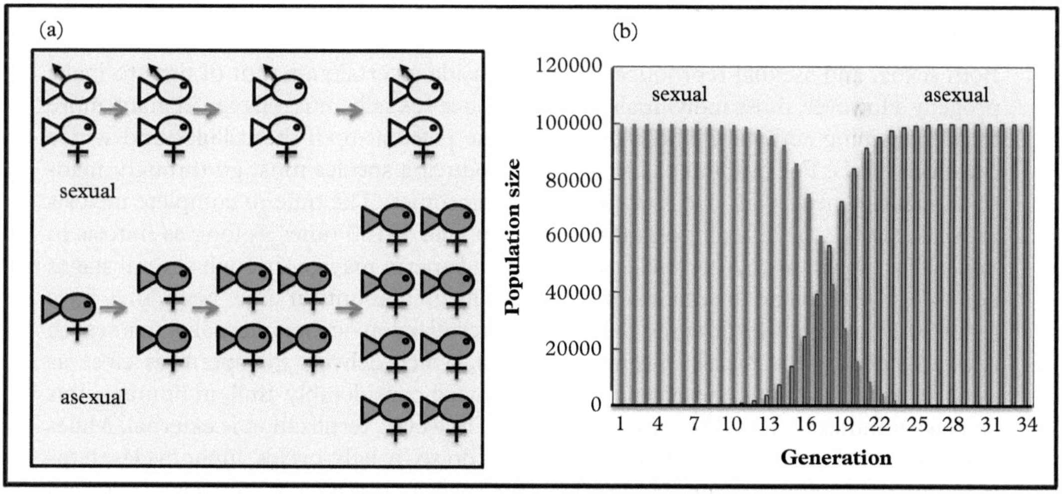

Figure 5.2 *(a) Using the same amount of energy, asexual females (shown in orange) can increase in a population at twice the rate of sexual individuals (shown in blue). This is referred to as the 'cost of males'. (b) In a population of 100,000 individuals, the number of asexual females will be greater than the number of sexual breeders by the eighteenth generation. By the thirty-third generation, there are fewer than two sexual individuals left.*

don't observe this outcome, something else must be happening to keep the sexual species around.

One of the problems is that the disadvantage isn't necessarily 'twofold'. A twofold advantage assumes that the males do nothing else to increase the reproductive performance of the female, like providing her with extra food or helping to raise the offspring. Any help the male gives (i.e. co-operation) reduces the advantage attributed to asexual reproducers. Conversely, any hindrance to the reproductive performance of females caused by males (i.e. conflict) results in a more significant increase in the reproductive advantage of asexuality. In other words, although there is still a disadvantage to sex, this can be reduced or increased by the behaviour of males.

5.4 The cost of mating

The two main costs of sexual reproduction are typically cited as the cost of meiosis and the cost of males. However, several anatomical features are developed, and many activities are undertaken only by those combining their genes with others. And sex is risky! Understanding all of these costs and risks (and there are no doubt ones I've missed) is necessary to understand the problem of sexual reproduction better. The following overview illustrates how much of the sexual organism's time and energy is devoted to reproduction and why it is so risky.

5.4.1 Time

Both sexual and asexual reproducers must set aside a certain amount of time to make progeny. However, those individuals who reproduce sexually must engage in many more time-consuming activities. These processes take place at both the cellular and whole organism levels. For example, all sexually reproducing species must go through meiosis to make gametes with one copy of the chromosomes. The time to complete meiosis is quite variable for different organisms but can take 5–100 times as long as mitosis in unicellular organisms.[5] After meiosis, the egg and sperm may go through several stages before fertilization and are often stored for significant amounts of time. Next, they have to find each other for fertilization to happen. If fertilization occurs internally, this search is often aided by the male's intromittent organ, which delivers the sperm as close as possible to the egg; thus, 'searching time' is reduced considerably. Still, in humans, this can take minutes to days. In many sea creatures, however, fertilization is external. Males and females release their gametes, stimulated to do so by light cycles, lunar cycles, temperature, or some other cue. Hopefully, sperm will find the eggs, but this can take a substantial amount of time, depending on the proximity of the eggs and sperm when they are released.

However, one of the most time-consuming activities often involves finding an appropriate member of the opposite sex with whom to mate. Mistakes here can be costly, especially for females. Such courtship rituals can take a significant amount of time in some species, while in others, they are negligible. Depending on various factors, puberty in girls usually begins between 8 and 14, while the average age of first pregnancy (in North America) is in the late 20s. In other words, human females in some societies are, on average, biologically capable of reproducing 15–20 years before becoming pregnant. Humans are unusual, however, as girls may enter puberty younger (for nutritional or hormonal reasons) and have babies later (almost certainly for sociological reasons). In most animals, females will begin to breed once reproductively mature. However, sexually reproducing individuals still need to find an appropriate mate, and courtship, when it occurs, can take a long time, lasting anywhere from a few seconds to a few months.

In *Drosophila* (the fruit fly or vinegar fly often used in genetics research and classes), males must court the female before she mates. She isn't about to combine her genes with just any male, and courtship gives her time to assess his qualities. He orients his body towards the female and then taps her with one of his forelegs. If she is receptive, he vibrates one of his wings to generate a courtship song. If everything is going well, he will lick her genitalia and attempt to mount her. Sometimes he's successful on his first attempt, but generally, he has to repeat the courtship ritual several times. The time requirement of courtship depends on the receptivity of the female but can take 10–15 minutes. Finally, he mounts the female and deposits his sperm, a process that takes another 10–20 minutes. This amount of time may seem insignificant, but remember, fruit flies don't live that long.

Different species of birds are well-known for their courtship displays. These displays are a way for females to assess both the genetic quality of the male as well as any other parental benefits he might contribute. Different males may sing, display plumage or other physical features, dance, preen, or feed their potential mate. They usually do more than one of these. Such courtship rituals may take a substantial amount of time but are necessary if mating is to occur, so males go through with it. The bowerbirds are a group of 20 species found in Australia and New Guinea. The males of these species have extensive courtship displays that involve building and decorating 'bowers' that help females decide if the male is of high quality. For example, Eastern Australia's male satin bowerbird (*Ptilonorhynchus violaceus*) builds a bower about a metre long and twice as high. He then decorates his construction and the adjacent area with several objects, usually blue or yellow (colours that differ in other species; Figure 5.3). These objects can be berries or flowers, or they can even be made by humans, such as ballpoint pens. The males may spend the better part of the year building and decorating these areas but don't reproduce if their displays are unacceptable to females.

5.4.2 Energy

Sexual reproduction requires energy for all these processes and behaviours. Cellular processes involved in gamete production and mating behaviours require energy that is no longer available for nonsexual activities. But individuals must also devote energy to the growth of primary sexual organs essential for reproduction, as well as secondary characteristics usually produced by one sex that may not be essential for breeding but increase their chance of getting a mate. These are often anatomical structures such as

Figure 5.3 *The satin bowerbird of Australia spends a substantial amount of time and energy in building and decorating bowers that will hopefully be met with the approval of one or more females.*
Credit: Imogen/Adobe Photo ID 282990242.

the antlers of male moose or the horns of the rhinoceros beetle but can be other energy-demanding activities such as vocalizations, behaviours, or the construction of bowers.

Male fiddler crabs (*Uca pugilator*) have one oversized claw that can make up 40% of their weight. They threaten rival males with this claw and wave it at potential mates to get their attention and show their size. They also drum on the sand to show their endurance. Females like males with large claws and fast drumming rates as they tell them something about the health of the male.[9] Additionally, and not too surprisingly, males with large claws tire more quickly during strenuous exercise,[10] suggesting that it is energetically demanding to carry them around.

How much energy does it take to grow such structures? Antlers consist of rapidly growing bony tissue that develops in male moose (and their close relatives) each year they are sexually mature, and weigh up to 36 kg (80 lb) or 5% of the body weight of older individuals. They start growing in the spring but are not used to court females until the fall 'rutting' season. During the winter, they fall off, and the process starts all over. To grow these antlers, males must have access to high-quality food. This food is essential because males may be energetically stressed as antlers grow, having just survived a long, cold winter. In ungulates (e.g. deer, moose), energy requirements are highest during the breeding season even though foraging rate decreases.[11]

Structures necessary to attract a mate may also increase the energy required to perform other nonbreeding activities. The moose's antlers must make walking difficult in dense forests, the male peacock's tail certainly increases the energy used while flying, and the large claw of the male fiddler crab may make it more challenging to feed. For this reason, individuals often shed such structures during the nonbreeding season.

5.4.3 Injury and death

Because courtship and even intercourse can be somewhat violent, sexual reproduction can result in injury and even death to participants. Yes, this is somewhat ironic, but reproduction isn't for the timid. Generally, when individuals compete with other individuals for opportunities to mate, they use nonphysical vocal or visual signals to prevent direct physical confrontation and possible injury. These signals are particularly important if the combatants have lethal weapons or poisons. However, sometimes these 'courtship rituals' aren't enough and physical altercations arise.

Northern elephant seals (*Mirounga angustirostris*) are among the largest seals and exhibit extreme sexual dimorphism, with males weighing up to 2000 kg and females less than 600 kg. Size dimorphism of this type generally indicates that males fight each other for access to females (Figure 5.4). In other words, it pays for males to be large since this makes them more intimidating and better fighters. In most cases, rival elephant seal males can settle disputes using visual and acoustic threats. However, males also engage in direct physical contests, and interactions can be pretty violent, often resulting in serious injury to one of the combatants. And while the loser often fails to pass on his genes, at least in that year, the most successful can reap successful benefits, siring up to 170 offspring over his life.[12]

The battles between males for access to females (or vice versa) or the resources needed by members of the opposite sex can get quite violent in many species. However, the act of having sex can also result in injury to one of the participants. Male sharks have two penis-like appendages called claspers at the base of their pelvic fins. During sex, males insert one of the claspers into the female to transfer sperm into her reproductive tract. However, grasping your mate in an aquatic environment can be challenging. This inconvenience is probably one of the reasons why so many sea creatures simply release their gametes and do not use internal fertilization. What makes it even more difficult is that a shark male has no appendages to use to hold the female steady. For that reason, he bites her in the neck or on one of her fins and curls his body around her, attempting to get into a position where he can insert one of his claspers. Not surprisingly, she finds this a little uncomfortable and tries to elude, or at least minimize his grasp. The skin of female sharks is thicker than that of males, presumably to reduce some of the damage that males might inflict during the mating process. Thus, sex in sharks can be relatively unpleasant and resembles a wrestling match between the two sexes.

Bed bug (*Cimex lectularius*) females have it particularly bad. A male doesn't fertilize a female through the reproductive tract but rather pierces the female's abdomen with his 'penis' to inject sperm near the eggs. Such a method is ominously, yet appropriately,

Figure 5.4 *The northern elephant seal is one of the most sexually dimorphic mammals. Two males fight for the right to mate with many females. The risks are substantial, but the rewards are potentially high.* Credit: Franco Folini/CC BY-SA 2.0.

referred to as '**traumatic fertilization**'. Fortunately, a female can use one male's sperm for up to six weeks since sex can result in serious wounds, lowering her longevity. Indeed, females generally leave the area relatively quickly, so any males can't pierce them again.[13] Such injuries that may occur to females during reproduction highlight the message that males and females often have different strategies for maximizing their fitness.

Females aren't the only ones who can incur injury in a reproductive encounter. Males of many spider species often face very dangerous situations when they try to have sex with females. The females are often much larger than the males and may attempt to eat them before, during, or after copulation. As you might imagine, this can be harmful to males, and they exhibit several behavioural traits that reduce their chance of being eaten. Golden orb weaver (*Nephila pilipes*) females can be up to 50 mm long, while males only grow to about 6 mm (Figure 5.5). Their small size allows males to go largely undetected when they clamber onto the web of females looking for sex. However, it's difficult to go unnoticed when trying to copulate, and the male often binds the female with fine silk strands to reduce her aggression.[14] This procedure provides males a safer environment and allows them to copulate repeatedly without fear of being cannibalized.

Not all the violence that results from sexual reproduction is experienced by adult males and females. In their attempts to fight off rivals, male northern elephant seals

Figure 5.5 *Two golden orb weaver spiders ready to mate. Females are often the larger sex, not because large size is needed for fighting, but because larger females can produce more eggs. However, this size difference comes with a cost for males. In this species, the smaller male often binds the female with threads to reduce the chance of being cannibalized.*

Credit: Percy Hendrich/Dreamstime Photo ID 167870781.

discussed earlier often fail to notice young seal pups in their way and may seriously injure or kill them unintentionally. Depending on the year, inadvertent activities by males can kill 13%–43% of pups before they reach the age of 28 days.[12] Before you say this doesn't seem very intelligent from an evolutionary viewpoint, these are pups sired by males the season before and so may be unrelated to this year's dominant males.

As noted previously, male lions may intentionally kill their potential mates' offspring. Lions occur in prides that usually consist of five to ten females and their offspring. Two or three males, usually brothers or cousins, will take over a pride when they can overcome the current male leaders. If the females have young offspring, the new male(s) will often kill the little ones.[15] This does two things. First, they eliminate any offspring that aren't genetically related to them. Second, the females will return to breeding condition and be more willing to mate with them. If they didn't kill off the previous male's offspring, current males might have to wait years to mate with the females.

Displays of any kind are often designed to make you more noticeable to members of your own or the opposite sex. Unfortunately, they also make you more conspicuous to other species, and some of these species may want to eat you. Túngara frogs (*Physalaemus pustulosus*), for example, can be found breeding in and around small pools in Central America. Males inflate their vocal sacs and produce one of two songs. The first is simple and designed to attract the female's attention. The second is more complex and encourages the females to mate. In other words, males that produce the most complex song when females are present are most likely to breed. The problem is that the complex vocalizations also leave physical traces as ripples in the water. Predatory bats can detect these ripples with their sonar, allowing them to home in on the frogs.[16] This can cause problems for the males, as you might expect.

The cost involved in wielding sexually dimorphic traits is often determined by examining increased mortality in sex that possesses them. In other words, individuals may sacrifice long-term survival for physical characteristics that make them better fighters or more attractive to females. Indeed, cross-species comparisons in birds and mammals suggest that the most dimorphic species also have the most biased adult sex ratios in favour of females.[17] In other words, males of these highly dimorphic species are dying faster than those of the less dimorphic species.

5.4.4 Disease

To engage in sexual reproduction, you often must get close to the other individual. Not always, of course. There are many species in which eggs are fertilized externally, and males and females stay apart. As you might imagine, determining the distance between the parents of particular offspring is often challenging in these species, and studies have generally focussed on the distance from parents where fertilization occurs. On the other hand, some organisms that employ external fertilization do come in close contact. In many fish, such as sticklebacks and largemouth bass, males build nests into which females deposit eggs. The males quickly follow behind the females and spread sperm on the eggs to fertilize them. Amphibian males clasp the females in a mating posture known as amplexus, and both sexes release their gametes close together; external fertilization in this group requires that males and females become rather intimate.

However, internal fertilization nearly always involves touching an individual of the opposite sex. Although most people would say this is good, it can be very unhealthy if pathogens are inadvertently transferred. Disease transmission can be highly detrimental to those who reproduce sexually and is a cost that isn't typically found among asexual breeders. It can reduce successful breeding during the reproductive season, render an individual infertile, and even result in death.

Not surprisingly, individuals who interact closely transmit infectious diseases at higher rates. For example, disease transmission tends to be more of a problem in larger than smaller animal colonies, such as in cliff swallows (*Petrochelidon pyrrhonota*).[18] The frequency of transmitted diseases should increase even further when individuals touch each other, as individuals often do when having sex. Even kissing another person can result in the transmission of 80 million bacteria.[19] Humans can also share a series of illnesses between partners during sex, including AIDS, hepatitis, human papillomavirus and genital warts, herpes, chlamydia, trichomoniasis, chancroid, gonorrhea, crabs, *Molluscum contagiosum*, syphilis, scabies, and lymphogranuloma venereum.

Although similar lists of sexually transmitted diseases (STDs) are challenging to find for other animals, we can expect that close contact between individuals during sexual intercourse results in the easier spread of disease. Indeed, some of the most common sexually transmitted infections in humans actually come from other animals. The viruses responsible for AIDS and herpes most likely came from chimpanzees, while we may have got gonorrhea from cattle. Brucellosis, or undulant fever, is a common bacterial disease transmitted during sexual contact among domestic livestock. In cows, it can result in spontaneous abortion, weak calves, or reduced milk yields. STDs are even important among invertebrates. Texas field cricket (*Gryllus texensis*) males, infested with the IIV-6 virus, tend to have sex more often, but the virus renders them infertile.[20] Ladybugs can also transfer a parasitic mite to their partner during sex, making them sterile.[21]

Chlamydia sp., a common bacterial STD in humans, is also found in birds, reptiles, and many mammals. *Chlamydia pecorum* is a leading cause of death in koalas (Figure 5.6) and is likely responsible for the decline of this species over the past 20 years.

Figure 5.6 *This koala female and her offspring risk getting chlamydia, a bacterial disease usually passed between individuals during sex but also passed from females to their offspring. Chlamydia is a serious problem in koalas and is probably responsible for the decline of many populations.*
Credit: Mathias Appel/CC0 1.0.

Some populations have a 100% infection rate. Infected individuals may incur blindness, pneumonia, urinary tract infections, infertility, and death. Although the disease is typically passed between adult males and females during sex, it can also be transmitted to offspring.

Females can assess the quality of a male if there is a correlation between a sexually selected trait, including the vigour of his displays and his health. These are 'honest signals'—they can't be faked. Weak males don't have the energy to produce them and must use their strength to find food and combat parasites. Little energy is left to develop the large antlers that the females find attractive, or court females by increasing the drumming rates on leaf litter. But I'm getting ahead of myself, and I'll have more to say about this and how potential mates are assessed in a later chapter.

5.5 Summary

Sexual reproduction is very costly to maintain. Beneficial allele combinations can be split up during recombination (the cost of meiosis), and half the population might contribute very little except their sperm (the cost of males). Besides these two often-cited costs of sex, sexual individuals can accrue other expenses. Organisms must develop and maintain specialized structures and behaviours for successful reproduction, which takes time and energy to produce. Injury and death are not uncommon occurrences while engaged in courtship or coitus. Furthermore, diseases can be passed between individuals, which is particularly important if fertilization is internal. Thus, there must be a real benefit for sexual reproduction to occur.

References

1. Crump J. Komodo dragon gives birth to three hatchlings without male partner. The Independent (UK Edition) [Internet]. 2020 Mar [cited 2022 Jul 5]. Available from: https://www.independent.co.uk/news/world/americas/komodo-dragon-gives-birth-no-male-partner-a9395671.html
2. Bernstein H, Bernstein C. Evolutionary origin and adaptive function of meiosis. In: Bernstein C, Bernstein H, editors. Meiosis [Internet] IntechOpen; 2013 [cited 2022 Oct 5]. Available from: https://doi.org/10.5772/56972
3. Williams GC. Sex and evolution. Princeton, NJ: Princeton University Press; 1975. p. 200.
4. Gene dilution was recognized as a nonissue in the mid-1970s by several biologists, including: Barash DP. 1976. What does sex really cost? Am Nat [Internet] 1976 Sep [cited 2022 Oct 5];110(975):894–897. Available from: https://doi.org/10.1086/283112; Dawkins R. The selfish gene. London: Flamingo; 1978. p. 240; Treisman M, Dawkins R. 'The cost of meiosis': Is there any? J Theor Biol [Internet] 1976 Dec [cited 2022 Oct 5];63(2):479–484. Available from: https://doi.org/10.1016/0022-5193(76)90047-3
5. Lehtonen J, Jennions MD, Kokko H. The many costs of sex. Trends Ecol Evol [Internet]. 2012 May [cited 2022 Oct 5];27(3):172–178. Available from: https://doi.org/10.1016/j.tree.2011.09.016

6. Otto S. The evolutionary enigma of sex. Am Nat [Internet]. 2009 Jul [cited 2022 Oct 6];174\(Suppl 1):S1–S14. Available from: https://doi.org/10.1086/599084
7. Maynard Smith J. The evolution of sex. Cambridge: Cambridge University Press; 1978. p. 242.
8. Barron J, Lawson T, Jensen P. Analysis of potential factors allowing coexistence in a sexual/asexual minnow complex. Oecologia [Internet]. 2016 Mar [cited 2022 Oct 7];180 (3):707–715. Available from: https://doi.org/10.1007/s00442-015-3522-0
9. Matsumasa M, Murai M, Christy JH. A low-cost sexual ornament reliably signals male condition in the fiddler crab *Uca beebei*. Anim Behav [Internet]. 2013 Jun [cited 2022 Oct 10];85(6):1335–1341. Available from: https://doi.org/10.1016/j.anbehav.2013.03.024
10. Allen BJ, Levinton JS. Costs of bearing a sexually selected ornamental weapon in a fiddler crab. Funct Ecol [Internet]. 2007 Jan [cited 2022 Oct 10];21(1):154–161. Available from: https://doi.org/10.1111/j.1365-2435.2006.01219.x
11. Parker KL, Barboza PS, Gillingham MP. Nutrition integrates environmental responses of ungulates. Funct Ecol [Internet]. 2009 Jan [cited 2022 Oct 10];23(1):57–69. Available from: https://doi.org/10.1111/j.1365-2435.2009.01528.
12. Le Boeuf BJ. Male-male competition and reproductive success in elephant seals. Am Zool [Internet]. 1974 Feb [cited 2022 Oct 12];14(1):163–176. Available from: https://www.jstor.org/stable/3881981
13. Stutt AD, Siva-Jothy MT. Traumatic insemination and sexual conflict in the bed bug *Cimex lectularius*. PNAS [Internet]. 2001 May [cited 2022 Oct 10];98(10):5683–5687. Available from: https://doi.org/10.1073/pnas.101440698
14. Zhang S, Kuntner M, Li D. Mate binding: male adaptation to sexual conflict in the golden orb-web spider (Nephilidae: *Nephila pilipes*). Anim Behav [Internet]. 2011 Dec [cited 2023 Feb 10];82:1299–1304. Available from: https://doi.org/10.1016/j.anbehav.2011.09.010
15. Packer C, Pusey AE. Adaptations of female lions to infanticide by incoming males. Am Nat [Internet]. 1983 May [cited 2022 Oct 12];121(5):716–728. Available from: https://userpages.umbc.edu/~hanson/Jane'sLions.pdf
16. Halfwerk W, Jones, PL, Taylor RC, Ryan MJ, Page RA. Risky ripples allow bats and frogs to eavesdrop on a multisensory sexual display. Science [Internet]. 2014 Jan [cited 2022 Oct 15];343(6169):413–416. https://doi.org/10.1126/science.1244812
17. Promislow DE. Costs of sexual selection in natural populations of mammals. Proc R Soc Lond B [Internet]. 1992 Mar [cited 2023 Oct 10];247:203–210. Available from: https://doi.org/10.1098/rspb.1992.0030
18. Brown CR, Brown MB. Coloniality in the cliff swallow: The effect of group size on social behaviour. Chicago: University of Chicago Press; 1996. p. 580.
19. Kort R, Caspers M, van de Graaf A, van Egmond W, Keijser B, Roeselers G. Shaping the oral microbiota through intimate kissing. Microbiome [Internet]. 2014 Nov [cited 2023 Jul 25];2:41. Available from: https://doi.org/10.1186/2049-2618-2-41
20. Adamo SA, Kovalko I, Easy RH, Stoltz D. A viral aphrodisiac in the cricket *Gryllus texensis*. J Exp Biol [Internet]. 2014 Jun [cited 2022 Oct 16];217 (11):1970–1976. Available from: https://doi.org/10.1242/jeb.103408
21. Rhule EL, Majerus ME, Jiggins FM, Ware RL. Potential role of the sexually transmitted mite *Coccipolipus hippodamiae* in controlling populations of the invasive ladybird *Harmonia axyridis*. Biol Control [Internet]. 2010 [cited 2022 Oct 16];53(2):243–247. Available from: https://doi.org/10.1016/j.biocontrol.2009.12.006

6

So Why Sex?

The evolution of sex is the hardest problem in evolutionary biology.

—John Maynard Smith

There's no question that sex is costly. Yet, it occurs in more than 99% of eukaryotic organisms. Why it occurs in such high frequency has been one of the most important questions in evolutionary biology. Obviously sexual reproduction has its advantages, or it wouldn't occur at all, and that's what we want to look at in this chapter. First, however, let's look at the two main problems with asexual organisms, as these may have contributed to the evolution of sex. I'll then present two general hypotheses that suggest why sexual reproduction is so common among eukaryotes. Finally, I'll look at experimental evidence from species that demonstrate both sexual and asexual reproductive modes to see if these shed light on these hypotheses.

6.1 The main problems with asexual reproduction

Think of an asexual individual, well-adapted to live in a particular habitat but born with a minor flaw, a genetic mutation. Nothing too serious, and it doesn't affect its survival ability, at least not drastically. When it reproduces, it clones little versions of itself, using fission, budding, or one of the other methods mentioned earlier. For this reason, all its offspring inherit this slightly deleterious mutation. Its lineage continues to breed for a few generations, individuals surviving despite this minor flaw when another harmful mutation occurs. Again, nothing fatal, but now they have two deleterious mutations. You see where I'm going—generally, individuals in an asexual population cannot eliminate mutations when they arise (unless the mutated segment of DNA returns to its original form, but this is pretty rare). Mutations simply accumulate over time.

To illustrate this, we can use our favourite reptile, the whiptail lizard. In the genus *Aspidoscelis*, one can find both sexual and asexual species of whiptail. A group of investigators sampled seven asexual and five sexual species from the Southern United States and found that the asexual species had a lower selective pressure in protein-coding regions of the DNA.[1] In plain language, random mutations resulted in subtle changes in proteins; while these proteins were still functional, they would be less efficient over time

The Evolution of Sex. Kevin Lee Teather, Oxford University Press. © Kevin Lee Teather (2024).
DOI: 10.1093/9780191994418.003.0006

than those found in the sexual line. Sex in other species was weeding out the less efficient mutations as they arose, while they were allowed to accumulate in asexual whiptails.

This scenario highlights a significant problem in asexual reproduction, recognized by Hermann Muller, a Nobel Prize winning geneticist. All the fittest individuals will end up with the slightly harmful mutation, and their lineage will eventually go extinct,[2] a process called 'Muller's ratchet' (Figure 6.1). Now, the second most fit individuals will become the fittest, and so on. These mutations could only be weeded out by recombination involving a homologous chromosome. Muller later suggested that the entire population was unlikely to go extinct through this accumulation of deleterious genes; instead, asexual lineages were more likely to suffer, and extinction would more likely occur, because of competition with sexual species.

Therefore, asexual species face two related problems. First, any mutations in the DNA, deleterious or not, are likely to persist. Once they appear, they are passed on to offspring, at least in the short term, as long as they don't significantly impact survival. Second, asexual breeders are relatively constant genetically; offspring tend to be clones of their parent. Sure, a certain amount of variability is introduced through lateral gene transfer. In other words, individuals can sometimes pick up new genetic material from others. However, the constancy in the genomes of asexual breeders is a substantial problem, as natural selection can't function without the options provided by variability. Sexual reproduction occurs when the genetic material of two individuals is combined to produce offspring that contain a mixture of the parental genomes. Thus, sexual reproducers can mask these deleterious mutations or damaged DNA (since they have a pair of homologous chromosomes) or eliminate them altogether (through natural selection operating on increased variation in the population). The following two sections provide hypotheses that focus on these two problems.

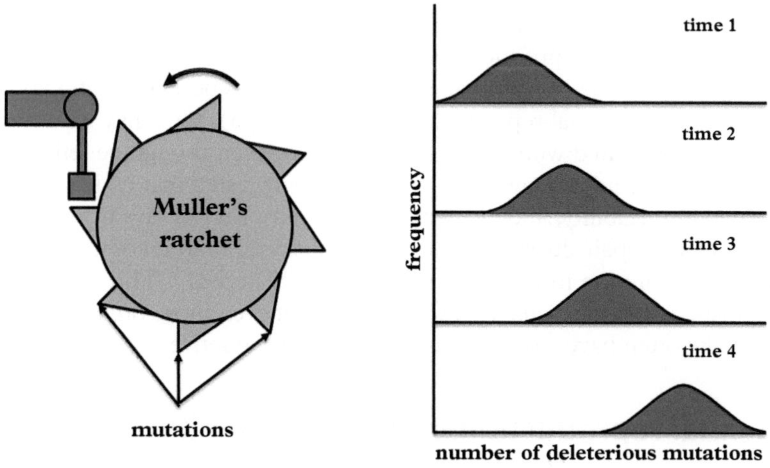

Figure 6.1 *Muller's ratchet describes the accumulation of mutations in asexual organisms. Mutations occur in each generation, and there is no way to eliminate them. Thus, the number of deleterious mutations increases over time.*

6.2 Sexual reproduction repairs DNA

It is clear that DNA suffers damage that can accumulate mutations over time. Many repair mechanisms have evolved in prokaryotes to deal with this damage. But homologous DNA strands, not present in individual prokaryotes, are often required to serve as templates to repair the genetic material. Many bacteria can address this by using one of the three methods of lateral gene transfer (transformation, conjugation, transduction) presented earlier. In Chapter 3, I discussed the possible evolution of meiosis from bacterial transformation, noting that the primary function of both processes was probably to repair DNA. One hypothesis concerning the evolution of sex, first proposed in the 1970s, is that sexual reproduction overcomes the damage and accumulation of mutations generated during each reproductive cycle.[3] This hypothesis simply extends the meiosis from transformation scenario, with DNA repair as a driving force behind meiotic evolution and sexual reproduction a necessary extension of meiosis.

As mentioned previously, DNA repair may have become increasingly important during the early evolution of eukaryotes. The incorporation of a bacterium that would become the mitochondrion resulted in increased aerobic respiration. In addition to the many external factors that might damage DNA, the organisms had to contend with harmful reactive oxygen species formed and released during respiration. Sexual reproduction probably evolved around the same time. The factors common to sexual reproduction[4] are paraphrased here: (i) Two different genomes come together in a shared cytoplasm; (ii) chromosomes containing the DNA become aligned; and (iii) an exchange of genetic material can occur between the chromosomes coming from different individuals. The latter two processes refer to meiosis, which occurs as part of every sexual cycle, highlighting how segments of chromosomes could be altered during Prophase I of meiosis.

The immediate advantage to meiotic sex may have been restoring the genetic material damaged by internally produced reactive oxygen species.[5] Although there are costs to both meiosis (e.g. the breaking apart of favourable gene combinations) and many other components of sexual reproduction (see Chapter 5), they may be outweighed by the advantages associated with DNA repair. Indeed, even though genetic variability is the most often stated advantage of sex, it has been suggested that (i) sex doesn't always increase genetic variability, and (ii) genetic variability isn't always a good thing.[6] It is argued that DNA repair during meiosis is more crucial than recombination to generate variation, as only the former directly affects cell survival.[7] This repair mechanism became indispensable to eukaryotes, so much so that even those organisms that usually breed asexually often have a sexual phase in their reproductive cycle.

6.3 Sexual reproduction increases variability

In populations where reproduction is asexual, there is a lack of variation; offspring are identical to their parents. While this may be good if the environment is relatively homogeneous, or similar from generation to generation, reduced variation is of little use if the environment is complex or changes temporally (Figure 6.2). In a metaphor to explain why variation was important in the evolution of sex, George Williams, a leading evolutionary biologist, likened it to a lottery, arguing that you'd be more likely to win if you bought a hundred tickets with different numbers than a hundred tickets with the same number.[8] For those with more of a sweet tooth than a gambling propensity, we could think of reproductive modes like an ice cream vendor. Would it be better to serve only vanilla or to have a range of different flavours?

6.3.1 Vicar of Bray hypothesis

Henry VIII, Edward VI, Mary I, and Elizabeth I ascended Britain's throne between 1509 and 1558. This was a turbulent time when loyalties to the church were met with rewards, while people in opposition could be burned at the stake. Although the Roman Catholic Church reigned supreme when Henry VIII first took power, it fell into disfavour partly because it wouldn't grant him a marriage annulment to Catherine of Aragon. He, therefore, took away the pope's powers and conferred authority over the church to himself. After Henry died, the Catholic Church gained strength again, followed by (again and existing until this day) the Church of England. The Vicar of Bray served under each monarch over a total period of about 50 years. While compatriots lost their positions (or lives), The Vicar of Bray managed to adapt to each religion, first being a papist, then Anglican, then a papist, and finally an Anglican again. His stated goal was to serve his flock (although a continuation to live may have also played into it), and if he had to switch churches for this to occur, so be it. However, his name became synonymous with somebody who changes his principles frequently, siding with the current belief system.

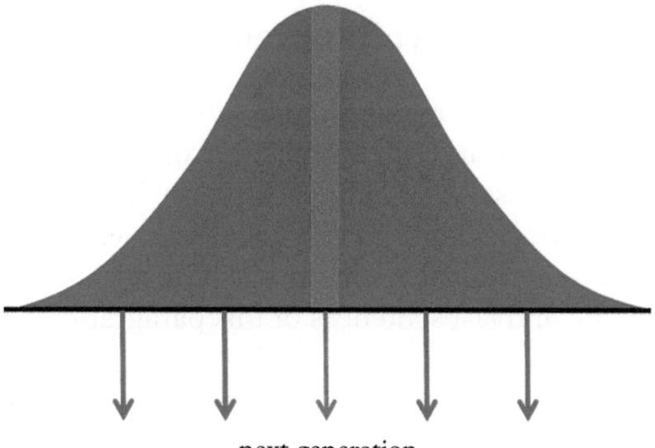

Figure 6.2 *One of the main advantages that sexual reproduction has over asexual reproduction is the ability to produce genetic variability. Asexual organisms, represented by the single blue line, clone themselves in each generation, making identical copies (for the most part) of themselves. Sexual organisms (in red) can produce offspring with various characteristics.*

The Vicar of Bray hypothesis (sometimes referred to as the Fisher-Muller model for much less interesting reasons) is based on one of the main advantages that sexual reproduction has over asexual reproduction—variability. The hypothesis states that offspring produced in a sexual population will be more genetically variable and thus be able to better respond to any environmental changes that are likely to occur over time. Natural selection would favour the best individuals or those containing the alleles that make them most suitable for that habitat. For this reason, populations of sexual reproducers would fare better than asexual ones in the long term.

'In the long term' is the problem here. The local environment is unlikely to change much from generation to generation; if the parent successfully lived there, her genetically identical offspring should also be successful. Why change a good thing? The Vicar of Bray hypothesis invokes group selection as a mechanism that permits sexual reproduction to evolve. In other words, sexual reproduction would benefit the entire population over the long term if the environment changed, but it would not favour individuals in the next generation. As discussed in the first chapter, 'group selection' is a phrase that makes most evolutionary biologists wary. 'Survival of the species' typically invokes 'group selection', and, although still used frequently in nature shows and magazine articles, is not readily accepted by evolutionary biologists. Sexual reproduction must be advantageous to individuals in each generation, not entire populations over the long term, if it is to evolve. When it comes to natural selection, it pays to be selfish.

The Vicar of Bray hypothesis wouldn't work mainly because asexual organisms would still have a short-term advantage, as we wouldn't expect the physical environment to change very much between generations. Over the long term, populations of sexual reproducers might have an advantage because they could respond better to long-term environmental changes; however, the asexual ones could produce more offspring in every generation until sufficient change occurred. And natural selection works in each generation—it has no foresight.

6.3.2 Spatial heterogeneity: Tangled bank hypothesis

Charles Darwin was criticized by many for taking God out of the equation when he discussed changes in organisms and biodiversity. He didn't say there was no God, only that he/she/it/they was unnecessary for natural selection. The last paragraph of his book on evolution through natural selection had very religious overtones, probably in response to this criticism. Indeed, there was no mention of a 'Creator' in his original treatise, and I provide you with the first and last sentences of this paragraph from the sixth edition of his book.[9]

> *It is interesting to contemplate a tangled bank, clothed with many plants of many kinds, with birds singing on the bushes, with various insects flitting about, and with worms crawling through the damp earth, and to reflect that these elaborately constructed forms, so different from each other, and dependent upon each other in so complex a manner, have all been produced by laws acting around us. ... There is grandeur in this view of life, with its several powers, having been originally breathed by the Creator into a few forms or into one; and that,*

> *whilst this planet has gone circling on according to the fixed law of gravity, from so simple a beginning endless forms most beautiful and most wonderful have been, and are being evolved.*

This eloquent description was partially usurped by Michael Ghiselin in 1974,[10] with his ideas later championed and expanded upon by Graham Bell when writing about the potential benefits of sexual reproduction.[11] Ghiselin called this idea the Tangled Bank hypothesis because it is expected that, in the habitat so nicely described by Darwin, there is intense competition for resources, including food, water, and space. If that environment is structurally diverse, the offspring arising from sexual reproduction would be expected to fare better for a couple of reasons. First, offspring that differ slightly from their parents or siblings should be better able than asexual clones to exploit a greater range of resources in a spatially heterogeneous habitat. Second, and related to the first, siblings slightly different from one another won't compete as intensely as identical ones.

If this hypothesis was correct, what species should exhibit sexual reproduction? We would expect sibling competition to be most intense when the density of individuals is greatest, as would occur when lots of offspring are competing for limited resources. Offspring that differ even slightly from one another, as would occur in sexual reproduction, would have a real advantage in these situations. However, this is not what we see in nature. Smaller species usually have more offspring that compete more intensely for access to resources. However, these are not the species in which we find sexual reproduction consistently. In fact, all the species of asexual breeders are small. At the same time, larger species, producing fewer and more widely spaced offspring are invariably sexual when it comes to reproducing. Bell himself later discarded this hypothesis in favour of the next one.

6.3.3 Temporal heterogeneity: Red Queen hypothesis

Two favourite books that served me well when I wanted to be distracted from my university studies were *Alice in Wonderland* and its sequel *Alice Through the Looking Glass*. In the second book, Alice visits the strange world of the 'Looking Glass', encountering all kinds of strange and memorable characters. One of these was the Red Queen, a living and human-sized version of the chess piece (Figure 6.3). The queen suddenly grabs Alice and begins to run very rapidly but never gets anywhere. Perplexed, Alice wonders why they never passed anything and, indeed, remained in the same place despite all their running. '*Now, here, you see*', says the Red Queen, '*it takes all the running you can do, to keep in the same place. If you want to get somewhere else, you must run at least twice as fast as that!*'

This phrase came to be used to describe the coevolutionary process between organisms that interact with one another. Leigh Van Valen initially used it to explain the evolutionary processes between predators and prey,[12] or parasites and their hosts. Suppose an adaptation in one organism evolves to make it more successful at its lifestyle (e.g. a toxin in prey to make them less palatable to enemies). In that case, there is selection pressure for a trait in an interacting organism that will enable it to deal with this

Figure 6.3 *The Red Queen told Alice she would have to run fast just to stay in the same place. This scenario has been used to explain coevolution between organisms and, particularly, that between pathogens and their hosts. As such, it is widely recognized as the most supported hypothesis to explain the evolution of sexual reproduction.*

Credit: John Tenniel 1871; public domain.

(e.g. the production of neutralizing gastric juices that will allow a predator to continue to eat the prey). Thus, there's an evolutionary arms race—species always must change because organisms around them are changing. This process stresses the biotic (living) community's importance as a significant selection pressure for evolutionary change; previous emphasis was often placed on temporal or spatial changes in an organism's environment's abiotic (nonliving) aspects.

Although the idea that parasites (or, more appropriately, pathogens) might be the force driving the evolution of sexual reproduction was first introduced by Levin,[13] the late William Hamilton is usually credited with developing this idea.[14] Hamilton was a British biologist who worked in both England and the United States and was an important evolutionary biologist in the twentieth century. Although many people might not wish to ponder how pathogens affect organisms (do you ever wonder what other organisms share your body?), parasitism is ubiquitous. Every multicellular creature is probably parasitized at some point in life by other organisms trying to get a meal or find a safe place to live. By living on or in its host, a parasite can use the host's resources to live out its life cycle, either entirely or partially, often progressing through many generations as they do so. Because of their potential impact on an organism's health, pathogens have been examined for their ability to be a selection pressure, especially regarding essential biological functions like sex.

Just how common are pathogens? Let's use humans as an example. We can look at how parasites have affected people over the years, and we can look at the number of parasites that live in or on us. These should give us a reasonable idea of how they can influence our evolution. Recent estimates suggest we have about 10 times as many cells of other organisms as our own in our bodies. Of course, these are not all parasites. Parasites, by definition, interfere with the physiological machinery of their host and cause some damage. This definition excludes many organisms that live in us (mainly bacteria) that cause no harm or even help us in some way. Still, it is estimated that humans are parasitized by at least 300 species of parasitic worms and 70 species of protozoans, about 90 of which are common.[15] This estimate omits the many bacterial and viral parasites whose numbers are difficult to ascertain but have affected human survival in the past and, given the recent COVID-19 pandemic, continue to do so. Pulmonary tuberculosis, diphtheria, cholera, leprosy, pertussis, tetanus, plague, gonorrhea, syphilis, and salmonellosis are just a few bacterial problems in humans. Viral diseases include herpes, chickenpox, smallpox, influenza, measles, rabies, hepatitis, autoimmune deficiency syndrome (AIDS), COVID, and many others. Most of these diseases will be familiar to you, and their high historical death rates make them substantial evolutionary selection pressures for humans.

Parasites are important because they typically breed very quickly and can adapt to a body that usually has a much longer generation time. The virus that is responsible for AIDS is a good example (Figure 6.4). Its reproduction is fast and error-prone, two traits that make it challenging to treat. After hijacking the machinery of a human host cell, one human immunodeficiency virus can generate billions of copies in 24 hours. Many of these copies will contain mutations that render them less susceptible or insusceptible to any drugs employed to combat them. These mutated lineages will then increase even as the drugs reduce the survival of the others.

Now let's return to sexual reproduction and see how it can combat parasitism. Sexual organisms produce a diverse array of young who have half the genes of each parent, mixed up so that they inherit some of the genetic characteristics of both their mother and father. Hamilton realized that parasites that have gone through numerous generations in their host are well-adapted to the habitat occurring in each parent. Suppose these well-adapted parasites are transmitted to the sexually produced offspring's body. In that case, they will be at a disadvantage, as the environment will be different from that of its parent. Think of sex like shuffling a deck of cards. If the cards are in the same order for an extended period, someone with a good memory could eventually memorize their order. This outcome would obviously be to their advantage, as they could determine the hands of all the players. After a while, the deck is shuffled, and this person no longer has the same advantage. Sex is a way of shuffling the deck in each generation.

Does the interaction between parasites and their hosts explain why sexual reproduction is so prevalent? Hamilton addressed this question using computer simulations.

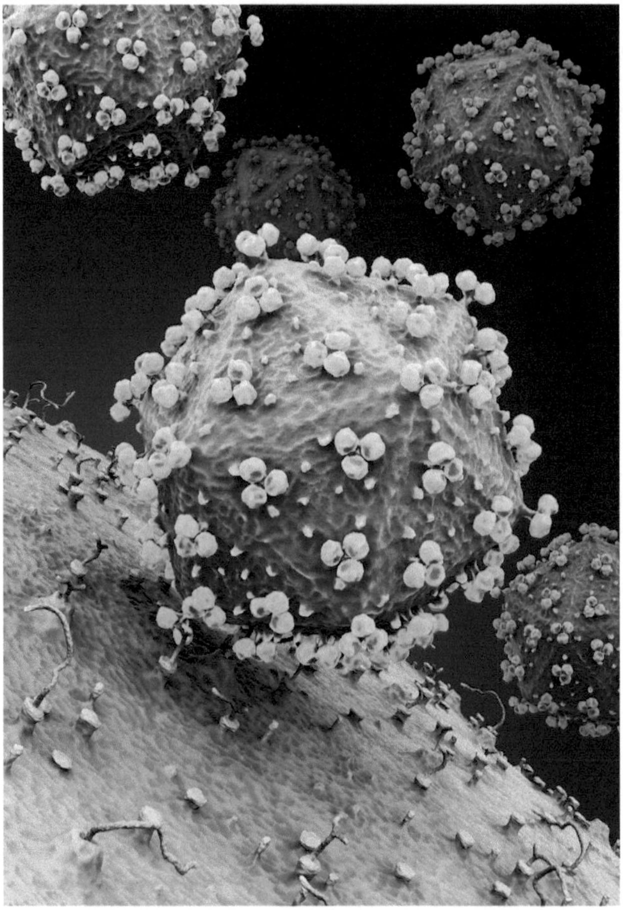

Figure 6.4 *The human immunodeficiency virus (HIV), seen in this drawing attacking a cell, can go through many generations during the lifetime of its host. Natural selection, therefore, has time to select an increasingly suitable viral genome that allows it to thrive in that habitat. If offspring were genetic clones of the parent, the pathogen's adaptation to the host would continue to improve. The variability resulting from sexual reproduction may be its main advantage; offspring that differ from their parents can better combat pathogens that have become adapted to their host.*
Credit: martynowi.cz/Shutterstock Photo ID 133011752.

He found, not surprisingly, that if asexual and sexual organisms lived in the same area and exploited the same resources, asexual organisms always won out; their populations increased over time, while those of the sexual organisms decreased. Asexual organisms clearly had an advantage over the short term in such situations. However, when pathogens were added to the artificial environment, the sexual organisms gained an advantage.[14]

6.4 A return to asexuality

Sexual reproduction was present or evolved in the early lineage of eukaryotes and occurs in nearly all of this group today. However, once sexual reproduction evolved, several species returned to asexual reproduction intermittently (facultatively) or all the time (obligately). Can these species tell us something about why sexual reproduction is generally favoured? A return to asexual breeding, at least facultatively, isn't that surprising, given the many problems with sex outlined previously. Asexual reproduction may be a valuable strategy to increase numbers over a brief period. Indeed, the true paradox of why sexual reproduction exists may be the prevalence of obligate sex rather than employing a combination of sexual and asexual reproduction.[16] Let's look at two organisms that exhibit facultative asexuality to see when sexual reproduction is used.

As I was tending my garden a while ago, I came across the scourge of the vegetable world—the wily aphid. These destructive pests suck the sap out of leaves and stems and can quickly spread if something isn't done about them. To be fair, I should note that only a few of these are agricultural pests—most are completely harmless, to us at least. But the fact that they can spread so rapidly makes them not only an insect to be feared but an interesting species in the study of reproduction. There are more than 4000 species of aphids, some that reproduce only asexually, most that use both asexual and sexual modes of reproduction, and none that use only sex to breed. By examining the species that are asexual breeders some of the time but reproduce sexually at other times, we might uncover some helpful information about the conditions under which both modes of breeding are most successful. This information should better allow us to discover the benefits of sex and might even provide insights into why it evolved.

Typically, aphids go through several parthenogenetic generations during the spring and summer and a single sexual one (although some species have given up this sex bit) in the fall. During the parthenogenetic stage, females produce only daughters, which emerge from the adults as tiny, well-formed aphids (Figure 6.5). Often, this development is so rapid that offspring are born already carrying their own progeny. For a brief time, this means that females will be sustaining two generations inside them—their daughters and granddaughters! A rapid increase in the number of aphids occurs this way throughout the summer, a time when environmental conditions are generally conducive to aphid growth and development. At the onset of cooler weather, the females parthenogenetically produce males and females. But wait a minute—how can a parthenogenetic female give birth to a male? It turns out that aphids can produce both sexes because of their genetic sex-determining system; males are identical to their mothers, except they have one fewer sex chromosome. Female aphids are XX, while males have only one X. Thus, mothers only have to hold back one of the X chromosomes to produce a male.

Females and males born near the end of the summer look different from the other aphids. One or both sexes are winged (females produced earlier in the summer never

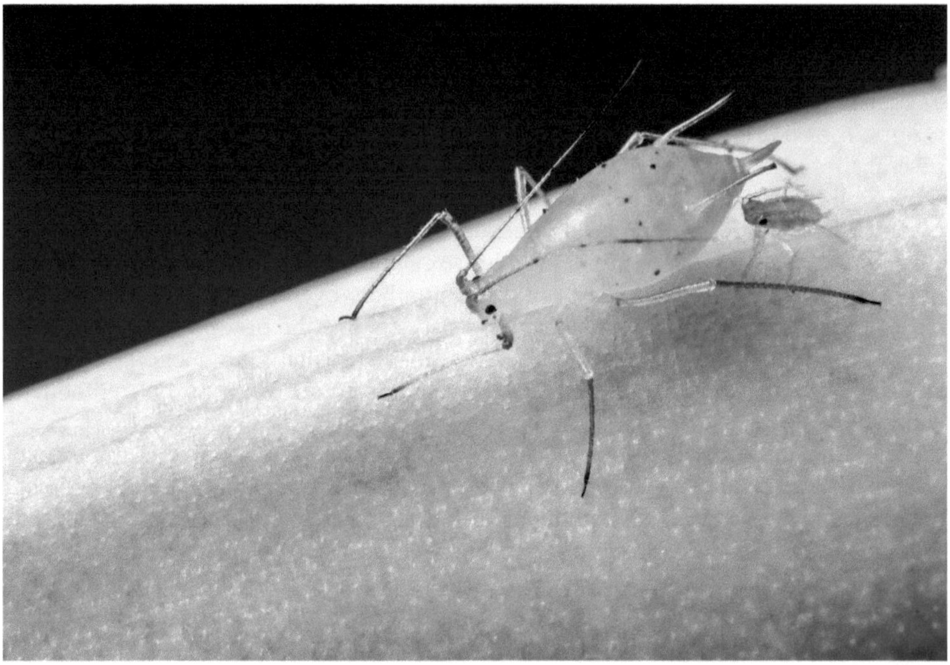

Figure 6.5 *There are over 5000 species of aphids, and most alternate between asexual (parthenogenesis) and sexual modes of reproduction. This female aphid is shown just after giving birth to a female clone of herself. The clone will already have her own daughters developing inside of her.*
Credit: Frances van der Merwe/Shutterstock Photo ID 407893780.

are), making it easier to colonize new areas. After these individuals engage in sex, the females manufacture one or a few eggs that are resistant to cold and can survive over winter, unhatched. In the spring, a sexually produced but parthenogenetic female emerges and begins the cycle again. Not surprisingly, the species that have a sexual phase occur in regions where overwintering in cold-resistant eggs makes sense; obligate asexual species are more likely to occur in areas where the winter tends to be mild so that they can continue their asexual lifestyle all year round.

What about other facultatively asexual breeders? More than 100 species of *Daphnia*, more commonly called water fleas, are small crustaceans living in freshwater environments. Their life cycle is very similar to cyclically asexual aphids, except that the females are oviparous, meaning they lay eggs rather than give birth to live young, during their asexual stage. After every moult, females produce a clutch of diploid eggs parthenogenetically. These eggs undergo development in her brood chamber, hatching after a day or so, and offspring emerge a few days later as miniature versions of their mother. After another week or so, these females lay their unfertilized eggs. Under good food conditions, the adult female continues to produce eggs every 3–4 days for the next couple of months, and then, just before winter sets in, she does something different. She generates diploid eggs that hatch into males. In addition, she now produces haploid eggs that the males fertilize. The production of these males occurs when environmental conditions get worse—the density of the population increases, and food becomes increasingly scarce. Shorter days and colder temperatures may also play a role. Sexually produced eggs differ from asexual ones, being larger, fewer in number, and enclosed in a protective casing. They will overwinter and then hatch as females in the spring when their parthenogenetic lifestyle begins again.

Both facultatively sexual aphids and water fleas breed asexually when environmental conditions are relatively good and sexually when conditions are less favourable or unpredictable. This observation would seem to favour the 'increased variability' role for sexual reproduction. Both groups of organisms would increase their numbers rapidly when conditions remained relatively stable and then breed sexually to generate the genetic variability favoured when conditions were more severe. However, DNA is damaged at a higher rate during stressful situations or when environmental conditions deteriorate, so the repair of DNA may also play a role. Therefore, both hypotheses could explain why sex is a more common overwinter strategy for these species.

What about those species that always reproduce asexually? There are two lineages of *Daphnia pulex*, the most common type of water flea—cyclically parthenogenetic and obligately parthenogenetic. The genotypes of the strains that always breed asexually have resulted from a meiosis-suppressing region of a chromosome, part of which was obtained through hybridization with a closely related species, *D. pulicaria*. However, without going into too much detail, these asexual lineages only have a lifespan of about 22 years.[18] Thus, the asexual lineages are destined for extinction; they burn brightly but are extinguished quickly, most likely because of the accumulation or expression of deleterious mutations. A similar scenario may occur in the obligately asexual aphids mentioned earlier. Unfortunately, it isn't clear how long obligate asexuals persist in the field for this group;[19] however, they may be relatively short-lived, much like the asexual lineages of *Daphnia*.

On the other hand, certain obligately asexual breeders are very successful and can persist over a long period. Bdelloid rotifers (Figure 6.6) are one of the few eukaryotes that reproduce asexually all the time, breaking off from their sexual relatives about 100 million years ago. They are somewhat famous to those studying sex as they are the most successful eukaryote in which no males, hermaphrodites, or meiosis has ever been found. These rotifers already possess substantial genetic diversity, having two to four copies of most alleles in a tetraploid genome. When stressed, bdelloid rotifers dehydrate and enter into a state of dormancy. It is during this state that their chromosomes shatter. When the organism revives, the genome is put back together in a jumbled mixture of genes, resembling recombination. Much of the genetic material found in bdelloids is derived from other organisms such as bacteria, fungi, or even plants.[20] The shuffling of genes and the incorporation of genetic material from other species not only generate the needed variability in this species, but also provide undamaged templates that function in increased repair to DNA.

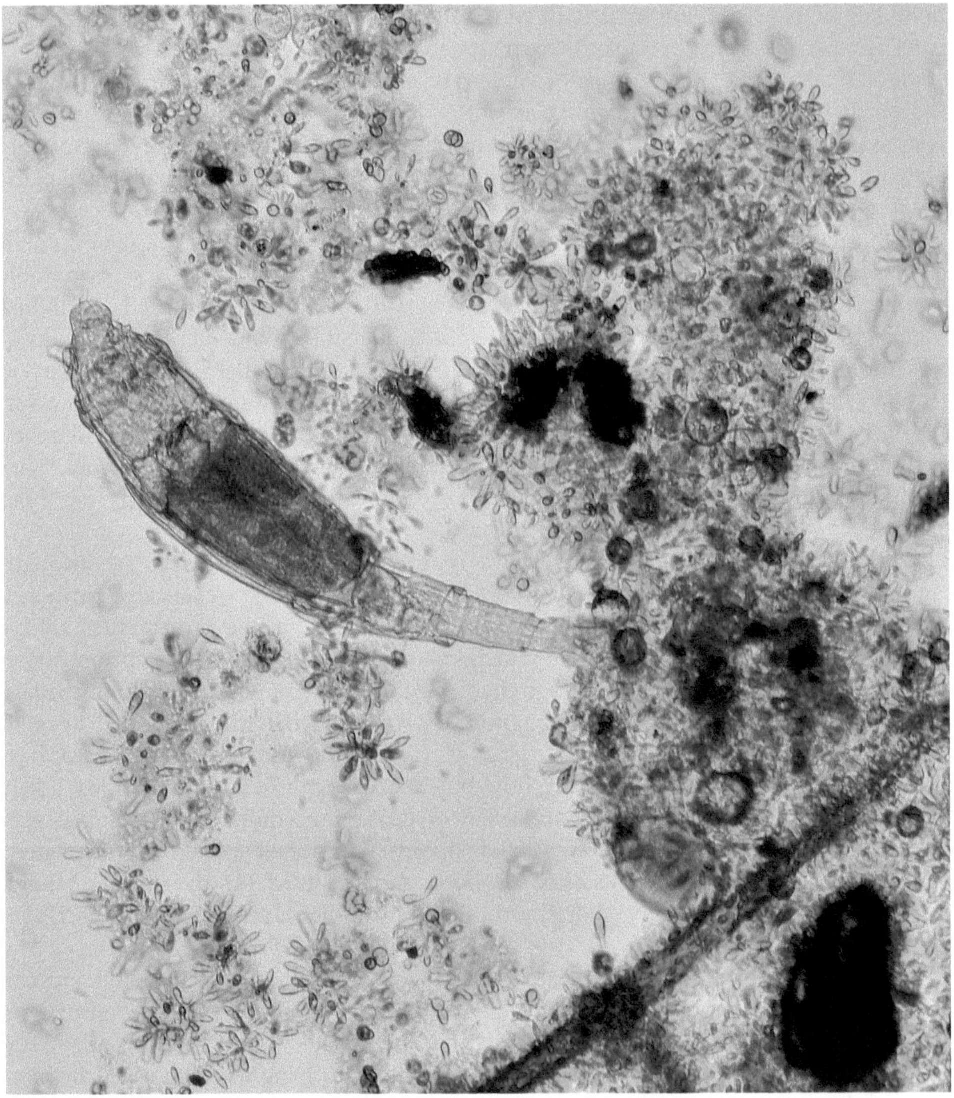

Figure 6.6 *Bdelloid rotifers are one of the few groups of eukaryotic organisms to revert to obligate asexuality. Shortly after the rotifer eggs hatch, the newly emerged females begin laying their own eggs. No males have ever been found.*

Credit: Eduardo/Adobe Photo ID 427727844.

6.5 Sympatric asexual and sexual populations

Lastly, we can examine examples of asexual and sexual lineages of the same or similar species coexisting in the same habitat. These are rare, but we do find them, and they've been used primarily to test predictions arising from the Red Queen hypothesis. Let's look at a couple of these groups and see if they provide evidence for or against sexual reproduction being used as a defence against pathogens.

Perhaps the earliest study on sympatric species having different reproductive modes to test predictions arising from the Red Queen hypothesis was conducted on topminnows in Mexico, studied by Robert Vrijenhoek, now retired from the Monterey Bay Aquarium Research Institute. These fish have a rather complicated mating system. *Poeciliopsis monacha* and *P. lucida* are sexual reproducers who can breed with each other because they are close relatives. More specifically, a female *monacha* can mate with a male *lucida*, producing all female offspring. Such females still require the sperm of male *lucida*, so they are not asexual. However, these female hybrids (*P. monacha-lucida*), by breeding with male *P. lucida*, then give birth to the next all-female generation, this time having three sets of chromosomes (triploid). These are asexual, and although they require sperm to get development started, don't incorporate any of the genetic material from the male. Yes, it's confusing. However, the important thing is that the closely related sexual and asexual species coexist in the same pools.

The Red Queen hypothesis predicts that sexually produced offspring will have lower parasite loads than asexual offspring. This is because the rarest genotypes (every individual created sexually is slightly different) will be less affected by parasites than the more common ones (each asexual individual is identical to its mother and sisters). One of the most common parasites of topminnows, sexual or asexual, is a type of trematode (*Uvulifer sp.*) that results in black spot disease. Confirming what was expected, researchers found that the common clonal asexually produced individuals were more heavily infected than the rarer sexually produced individuals.[21] The one exception to this occurred after a major drought; it was then found that the sexual females had more trematode cysts than their asexual cohabitants, seemingly contradicting the Red Queen hypothesis. Upon closer examination, however, it was observed that both the sexual and asexual populations of topminnows were decimated as water levels got lower, and the sexual species, now heavily inbred, had a much more common genotype. When genetically variable sexual females were transplanted into this pool, parasite loads predictably decreased in the sexual lineage.

In the most extensive series of studies, Curtis Lively and his colleagues have examined populations of the tiny New Zealand freshwater mud snail *Potamopyrgus antipodarum* to test the relationship between sex and parasitism. This is a rather odd snail in which individuals can exist in a sexual (diploid) or asexual (triploid) form. Both of these lineages seem to perform equally as well, with asexual and sexual females maturing at the same rate and both producing the same number of young. At least 20 species of trematodes parasitize New Zealand mud snails. One, though, seems to be both common and particularly harmful—a sterilizing trematode called *Microphallus sp.* The snails

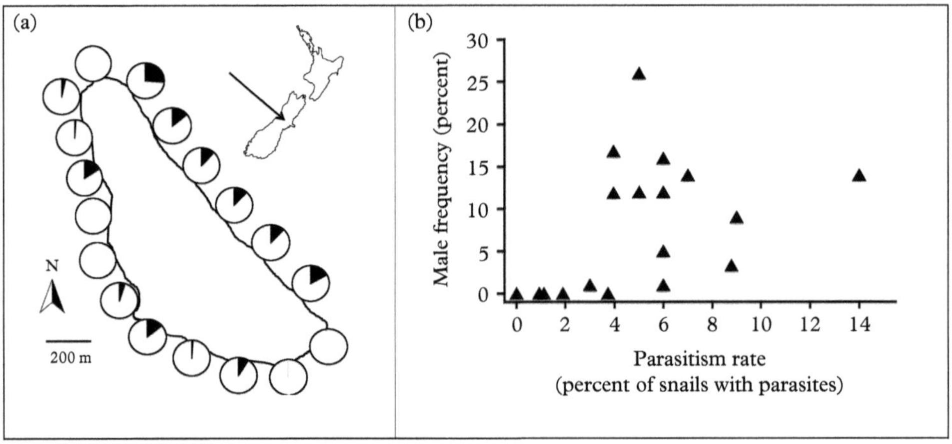

Figure 6.7 *a) Sampling sites and the relative frequency of male mud snails at different locations in a New Zealand Lake. The percentage of males in the population is a reasonable estimate of the level of sexual reproduction. The 18 sample sites are about 200 m apart, and the relative frequencies of males are based on approximately 100 individuals. The location of the lake (Grasmere) is shown on the inset of New Zealand. b) Sexual reproduction increases (males are more numerous) as parasitism by sterilizing trematodes increases.*

Credit: McKone et al.;[24] permission of the *New Zealand Journal of Ecology*.

are an intermediate host of these trematodes. The adults live in ducks, and their eggs are found in the duck faeces. Snails, for whatever reason, love duck faeces and ingest the eggs. When the eggs hatch in the snails, the trematode rapidly reproduces (asexually), resulting in the sterilization of both sexes of snails when they occur.

Let's look at three general hypotheses concerning the predicted relationship between parasitism and sex that have been tested using mud snails. First, where asexual and sexual forms occur together, the sexual individuals should have a lower parasite load, as with the topminnows described previously. Second, where the two types occur separately, we should find more sexually reproducing snails in regions with more parasitism, as their variable genotype will provide sufficient protection against the parasites. Third, as parasitism increases, individuals would be most successful if they could increase the variability of their offspring.

Investigators examined parasite loads of sexual and asexual females in a lake on South Island, New Zealand.[22] They looked at the frequency of both types of reproduction at four sites in Lake Alexandrina over five years. Because asexual reproducers are triploid, they contain more DNA than diploid sexual reproducers, and the two lineages can be identified by the amount of genetic material they have. It was observed that sexual females carried lower parasite loads than asexual ones, supporting the first prediction. Sexually reproducing females had a definite edge when it came to being infected by parasites, at least the sterilizing nematode.

To determine whether sexual mud snails are more common where parasitism rates are high, one can look at broad and narrow geographic areas. In this case, the frequency of sexual reproduction was estimated by the number of males present. No males, means no sex. King and Lively[23] found that sexual forms were more common where trematode infections were highest in a survey of 26 New Zealand streams. In a narrower geographic region, investigators collected 100 snails from 18 sites around a small (0.63 km^2) New Zealand lake[24] (Figure 6.7a). Again, it was assumed that the frequency of sexual reproduction was correlated with the number of males in the sample. Sterilizing trematode infection rates for the 1800 snails ranged from 0 to 14%. Most significantly, the percentage of males was positively correlated with the number of parasites (Figure 6.7b). Thus, whether we look at broad or narrow geographic regions, sex seems to occur more frequently where parasitism rates are high. Sexual individuals contain the variability needed to combat the parasitic trematodes effectively.

Lastly, how does one combat ever-changing parasitism rates? One way to do this is to vary the genotypes of your young. By making them more variable, you don't give the parasites time to adapt to one type. Of course, an asexual individual can't just switch to being sexually reproductive to increase the variability of its progeny. But if she is already a sexual reproducer, some things can be done. One possibility is to mate with many different males. Different males have different genes, so her offspring would, by virtue of having many fathers, have many different genes. Sexual females, indeed, became more promiscuous when exposed to varying numbers of trematode eggs (to signify different levels of parasitism).[25] Thus, increased parasitism drove up the number of sexual partners each female had.

The Red Queen hypothesis offers both short-term and long-term benefits to sex as ways to combat the ubiquitous parasitism of most organisms. I have looked at the species belonging to two groups, topminnows and snails, because they can exist in both asexual and sexual forms in the same environment, making it relatively easy to test predictions arising from the Red Queen hypothesis. But I don't want to leave you with the impression that these are the only supportive studies. Substantial recent evidence from various organisms aligns with this hypothesis,[26] although many investigations have found no support.[27] However, parasitism may not be the only answer. In any situation where an organism finds itself in an unpredictable environment, sexual reproduction may be beneficial. Parasitism, because it probably occurs in all multicellular organisms, simply provides a general benefit to sex that can be applied to most species.

6.6 Summary

Despite the many costs of sexual reproduction, nearly all eukaryotes use it to transmit their genes to the next generation. Although a few have returned to asexuality for reproducing, many still use sexual reproduction at some point in their reproductive cycle. Only a few species have converted to obligate asexuality. Thus, sexual reproduction must provide a distinct advantage for eukaryotes. Hypotheses that outline the selective advantage of sex fall into two camps. First, sexual reproduction provides the ability, during meiosis, to repair damaged DNA and get rid of harmful mutations whose numbers increased in eukaryotes because of aerobic respiration. Second, sexual reproduction provides the increased variation natural selection needs to adapt to local conditions.

Current evidence favours the Red Queen hypothesis, which promotes sex as a method to minimize the damage caused by pathogens. Parasites typically have many generations that enable them to adapt to the microhabitat provided by the host's body. By varying the genotypes of their offspring, sexual reproducers can decrease their suitability as a habitat for these pathogens. Field studies tend to support predictions arising from the Red Queen hypothesis, although more research is required to shed further light on the problem. On the other hand, repairing DNA is an obvious advantage to sexual reproduction and must be considered in any discussion regarding its evolution. Even though sex might provide a benefit against parasitism today, it isn't clear if variability was the initial selective force. It is possible that sexual reproduction was initially advantageous because of the benefits conveyed by repairing damaged DNA and was later maintained because of the benefits of increased variability. Unfortunately, this topic will unlikely be solved to everyone's satisfaction soon.

References

1. Maldonado JA, Firneno T Jr, Hall AS, Fujita MK. Parthenogenesis doubles the rate of amino acid substitution in whiptail mitochondria. Evolution [Internet]. 2022 Jul [cited 2023 Jan 10];76(7):1434–1442 Available from: https://doi.org/10.1111/evo.14509
2. Muller HJ. Some genetic aspects of sex. Am Nat. 1932 Mar–Apr;66(703):118–138.
3. Bernstein H. Germline recombination may be primarily a manifestation of DNA repair processes. J Theor Biol. 1977 Nov;69(2):371–380.
4. Bernstein H, Byers GS, Michod RE. Evolution of sexual reproduction: Importance of DNA repair, complementation, and variation. Am Nat. 1981 Apr;117:537–549
5. Hörandl E, Speijer D. How oxygen gave rise to eukaryotic sex. Proc Biol Sci [Internet]. 2018 Feb [cited 2023 Jan 16];285(1872);20172706. Available from: https://doi.org/10.1098/rspb.2017.2706
6. Otto SP. The evolutionary enigma of sex. Am Nat [Internet]. 2009 Jul [cited 2023 Jan 18];174:S1–S14. Available from: https://www.jstor.org/stable/10.1086/599084
7. Mirzaghaderi G, Hörandl E. The evolution of meiotic sex and its alternatives. Proc R Soc B [Internet]. 2016 [cited 2023 Jan 18];283:20161221. Available from: http://dx.doi.org/10.1098/rspb.2016.1221
8. Williams GC. Sex and evolution. Princeton, NJ: Princeton University Press; 1975. p. 210.
9. Darwin C. The origin of species by means of natural selection or the preservation of favoured races in the struggle for life. 6th ed. [Online]. 1873 [1st ed. published 1859]. Available from: https://www.literature.org/authors/darwin-charles/the-origin-of-species-6th-edition/preface.html
10. Ghiselein MT. The economy of nature and the evolution of sex. Berkley: University of California Press; 1974. p. 346.
11. Bell G. The masterpiece of nature: The evolution and genetics of sexuality. Berkley: University of California Press; 1982. p. 378.
12. Van Valen L. A new evolutionary law. Evolutionary Theory [Internet]. 1973 [cited 2022 Dec 1];1(1):1–30. Available from: https://www.mn.uio.no/cees/english/services/van-valen/

evolutionary-theory/volume-1/vol-1-no-1-pages-1-30-l-van-valen-a-new-evolutionary-law.pdf

13. Levin DA. Pest pressure and recombination systems in plants. Am Nat. 1975;109:437–451.
14. Hamilton WD. Sex vs. non-sex vs. parasite. Oikos. 1980;35:282–290.
15. Cox FE. History of human parasitology. Clin Microbiol Rev [Internet]. 2002 [cited 2023 Jan 14];15(4):595–612. Available from: https://doi.org/10.1128/CMR.15.4.595-612.2002
16. Burke NW, Bonduriansky R. The paradox of obligate sex: The roles of sexual conflict and mate scarcity in transitions to facultative and obligate asexuality. Evol Biol [Internet]. 2019 Aug [cited 2023 Jan 17];32(11):1230–1241. Available from: https://doi.org/10.1111/jeb.13523
17. Bernstein H, Bernstein C. Evolutionary origin and adaptive function of meiosis. In: Bernstein C, Bernstein H, editors. Meiosis [Internet]. IntechOpen; 2013 Apr [cited 2022 Sep 20]. Available from: https://doi.org/10.5772/56972 DOI 10.5772/56,972
18. Tucker AE, Ackerman MS, Eads BD, Xu S, Lynch, M. Population-genomic insights into the evolutionary origin and fate of obligately asexual *Daphnia pulex*. Proc Natl Acad Sci USA [Internet]. 2013 Jul [cited 2023 Jan 17];110(39):15740–15,745. Available from: https://doi.org/10.1073/pnas.1313388110
19. Loxdale HD, Balog A, Biron DG. Aphids in focus: unravelling their complex ecology and evolution using genetic and molecular approaches. Biol J Linn Soc Lond [Internet]. 2020 Mar [cited 2022 Dec 1];129(3):507–531. Available from: https://doi.org/10.1093/biolinnean/blz194.
20. Hecox-Lea BJ, Welch DB. Evolutionary diversity and novelty of DNA repair genes in asexual Bdelloid rotifers. BMC Evol Biol [Internet]. 2018 Nov [cited 2023 Jan 16];18:177. Available from: https://doi.org/10.1186/s12862-018-1288-9
21. Lively CM, Craddock C, Vrijenhoek RC. Red Queen hypothesis supported by parasitism in sexual and clonal fish. Nature [Internet]. 1990 Apr [cited 2023 14 Jan];344(6269):864–866. Available from: https://www.mbari.org/wp-content/uploads/2016/01/Lively_EtAl_1990.pdf
22. Vergara D, Jokela J, Lively CM. Infection dynamics in coexisting sexual and asexual host populations: Support for the Red Queen hypothesis. Am Nat [Internet]. 2014 Aug [cited 2023 Jan 12];Suppl 1, S22–S30. Available from: https://doi.org/10.1086/676886
23. King KC, Lively CM. Geographic variation in sterilizing parasite species and the Red Queen. Oikos [Internet]. 2009 Aug [cited 2023 Jan 12];118(9):1416–1420. Available from: https://doi.org/10.1111/j.1600-0706.2009.17476.x
24. McKone MJ, Gibson AK, Cook D, Freymiller LA, Mishkind D, Quinlan A, et al. Fine-scale association between parasites and sex in *Potamopyrgus antipodarum* within a New Zealand lake. N Z J Ecol [Internet]. 2016 [cited 2023 Jan 17];40(3):330–333. Available from: https://www.jstor.org/stable/26198766
25. Soper DM, King KC, Vergara D, Lively CM. Exposure to parasites increases promiscuity in a freshwater snail. Biol Lett [Internet]. 2014 Apr [cited 2023 Jan 15];10(4). Available from: https://doi.org/10.1098/rsbl.2013.1091
26. Martín-Peciña M, Osuna-Mascaró C. Digest: The Red Queen hypothesis demonstrated by the *Daphnia-Caullerya* host-parasite system. Evolution [Internet]. 2018 Jan [cited 2023 Aug 23];72:715–716. Available from: https://doi.org/10.1111/evo.13439
27. Arakelyan M, Harutyunyan T, Aghayan SA, Carretero A. Infection of parthenogenetic lizards by blood parasites does not support the 'Red Queen hypothesis' but reveals the costs of sex. Zoology [Internet]. 2019 Oct [cited 2023 Aug 23];136:125709. Available from: https://doi.org/10.1016/j.zool.2019.125709

7

Determining an Individual's Biological Sex

7.1 Adaptive sex-ratio manipulation

Males and females often have quite different ways of maximizing their fitness, but does it matter whether they produce sons or daughters? Barring certain exceptions, the number of males and females should be similar at the population level.[1] There may be theoretical reasons to expect minor deviations from a 1:1 sex ratio (i.e. if one sex is more costly to produce, it should be found in slightly lower numbers). Particular parents, however, might be expected to have more of one sex than the other. Trivers and Willard[2] suggested that parents, particularly mothers, who could skew the sex ratio of their offspring towards the sex that would benefit them more would have an evolutionary advantage. Let's assume that females of high quality have offspring of high quality. Given this, having sons may be beneficial as we expect that high-quality sons to have many offspring when they are reproductively mature. On the other hand, the reproductive output of females isn't as variable as that of males; the number of expensive eggs confines an upper limit to the number of offspring that can be produced (Figure 7.1). In some cases, females could still benefit preferentially from additional resources that high-quality parents might provide. For example, in many primate societies, females inherit their mother's social status; if mothers have a high rank, they might be better off producing female offspring if their daughters gain a better social position.

Many studies have examined whether parents can influence the sex ratio of young, with mixed results. Most often, female birds are the heterogametic sex and presumably, are in a better position to control the sex of their offspring. However, tests of Trivers and Willard's ideas have also been examined using mammals and insects. There's a growing body of evidence in some species of birds the sex ratio of the young can be manipulated in a way predicted by Trivers and Willard, while in others, such control has not been found or remains ambiguous.[3] In other groups, such as mammals, the evidence is less clear.[4] So, while the mothers in some species seem able to influence the sex ratio of their progeny, they are unlikely to control it entirely. However, mothers in other species can. We have seen that in aphids and water fleas, both facultative sexual breeders, mothers produce females all summer but produce males at the end of the season. This

The Evolution of Sex. Kevin Lee Teather, Oxford University Press. © Kevin Lee Teather (2024).
DOI: 10.1093/9780191994418.003.0007

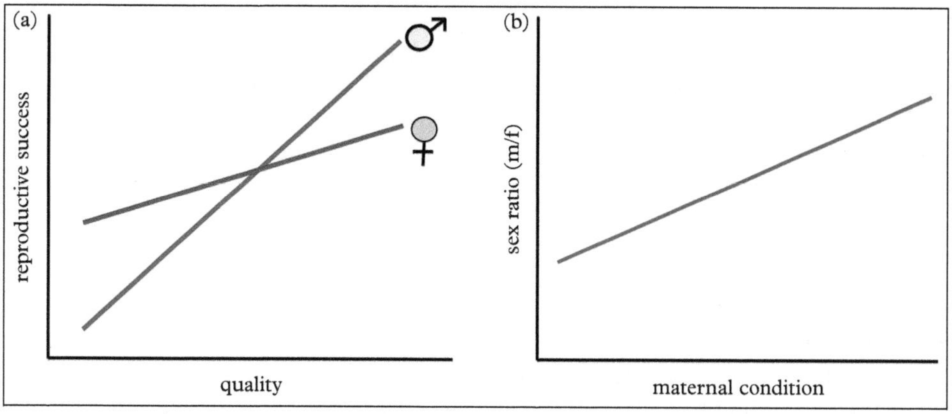

Figure 7.1 *A simple overview of the Trivers-Willard model concerning sex allocation. (a) The relative reproductive success of males tends to be more variable than that of females. Thus, high-quality males are expected to sire more offspring than high-quality females. Just the opposite is true for low-quality individuals. (b) Because of this difference in reproductive variability, females capable of producing high-quality sons should have a greater reproductive payback. For this reason, females in good condition should produce more males than females.*

reproductive cycle permits the next generation to overwinter in sexually created eggs. We see similar control over the sex of progeny in several of the haplodiploid insects described in Section 7.2.2.

Do situations exist where mothers prefer offspring of one sex and fathers the other? We typically think that high-quality males and females will produce high-quality offspring, regardless of their progeny's sex. However, as will be clear when we look at the conflict between males and females in the next chapter, the same genes that make a high-quality male may be detrimental to females and vice versa. This situation can be caused by a process called intralocus conflict, in which the same genes code for the same trait in both males and females, but the fitness optimum for that trait differs between the sexes. Therefore, males having high-quality genotypes would favour sons, while high-quality females would prefer daughters. In such cases, there is the potential for conflict between parents in the sex ratio of their offspring.

There is little evidence of parental conflict in the optimal sex ratio of progeny, although it has not been examined extensively. Contrary to the Trivers-Willard hypothesis, we would predict that females in good condition would typically have fewer male offspring, not more. On the other hand, low-quality females should have higher numbers of male offspring. There is some evidence of this from broad-horned flour beetles (*Gnatocerus cornutus*), where high-quality females produce more daughters, and low-quality females have more sons.[5] On the other hand, we would expect high-quality males to contribute more to sons. In the case of broad-horned flour beetles, high-quality reproductive males sire low-quality female offspring, providing some evidence of a conflict between the parents.

The evolutionary aspects of sex determination are complex, and as I'm unlikely to do justice to the topic in this short space, I'll focus mainly on the proximate mechanisms. Before anything was known about sex chromosomes, most believed that the environment established the sex of an individual. As early as Aristotle, temperature was thought to play a role in determining whether an animal was to become male or female. He believed that the temperature of the parents influenced the sex of the offspring, with males being hot and females being cold. In fact, males who wanted male heirs were advised to conceive their offspring in the summer. If the temperature of one of the parents overwhelmed that of the other, an individual of that sex would develop. Aristotle was close. For many reptiles, the temperature at which the egg develops determines its sex. But hot temperatures don't always lead to males.

We now have a much better understanding of how sex is regulated. Males and females are defined according to their gamete size, but what factor or factors determine which sex an individual becomes? Most people think that the biological sex of organisms is under genetic control, with males having X and Y chromosomes and females having two X chromosomes. But this is only true in some organisms; it just happens that humans are one of them. We now know two very different ways that dictate whether a male or female will develop. The first is the most familiar to us and occurs when the genetic makeup of the zygote dictates the zygote's sex; this is referred to as 'genetic sex determination' (GSD). The other method happens after conception. In this case, some aspect of the environment (usually temperature) determines the sex of the developing individual. This process is called 'environmental sex determination' (ESD). Let's first review how sex is determined before addressing whether a particular sex is advantageous in some situations. I'll then briefly look at species in which an individual can harbour both sexes, either simultaneously or sequentially. For a more complete discussion of sex-determining systems, Leonard provides an excellent review.[6]

7.2 Genetic sex determination

7.2.1 XX-XY and ZZ-ZW Systems

Genetic sex determination generally falls into two types: XX-XY, when females are the homogametic sex (have two X chromosomes), and ZZ-ZW, when males are the homogametic sex (Figure 7.2). While these are the most common, regardless of whether we look at vertebrates or invertebrates, there's a great deal of variation in how sex is determined genetically. Let's begin with the XX-XY system since that's the one with which we're most familiar.

In humans and most other placental mammals, males produce gametes containing either an X or Y chromosome, while females invariably produce gametes having only X chromosomes. If a Y-containing sperm fertilizes an egg, the resulting zygote will be male; if the sperm carries an X chromosome, the embryo develops as a female. Thus, the males of mammalian species are often said to determine sex even though the female actually gives birth to the young. Besides placental mammals, the XX-XY sex-determining

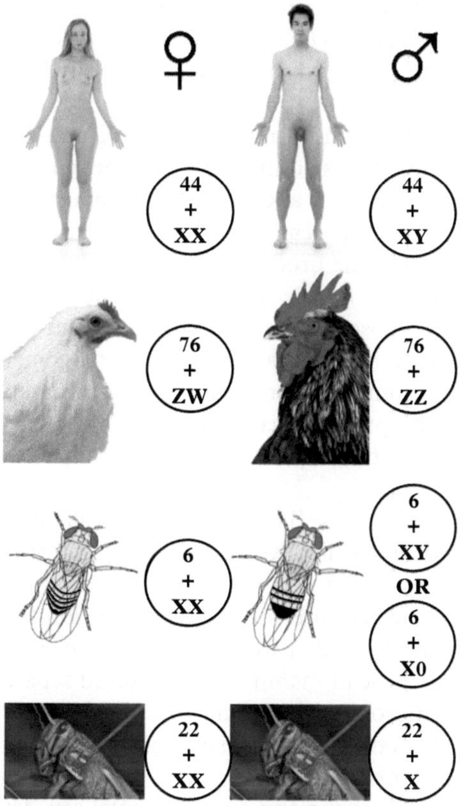

Figure 7.2 *A few of the genetic sex-determining mechanisms found in animals. In each case, the total number of autosomes plus sex chromosomes are provided.*
Credit: CFCF/CC BY-SA 3.0.

system also occurs in many other vertebrates (some fish, amphibians, and reptiles) and many insects. Not surprisingly, how sexes are determined in many species has not yet been discovered.

The African pygmy mouse (*Mus minutoides*) is an exception to the pattern generally observed in placental mammals. It possesses a modified version of the X chromosome (X'), which is feminizing. Thus, females can be XX, XX', or X'Y, with the X' overriding the (usually) male-determining Y chromosome. Males still have the XY genotype. The monotreme mammals, which include the platypus and four species of echidnas (spiny anteaters), are the most ancient group of mammals, falling between the modern placental mammals and reptiles on the evolutionary tree. Indeed, they express many anatomical and physiological traits that seem to be a mixture of both groups. The platypus (*Ornithorhynchus anatinus*) has five sex chromosome pairs; males are XXXXX-YYYYY. However, all the X's and Y's together act like individual chromosomes, attaching to each other and remaining like this during the production of sperm or eggs. The X chromosome of the platypus also seems to be intermediate between the two groups, originating from the Z sex chromosomes of birds at one end while similar to the X chromosome of mammals at the other.

Compare this to fruit flies, another group with X and Y chromosomes. In these species, the sex of an individual is controlled by the proportion of female-determining genes present on the X chromosome and male-determining genes on the autosomal (neither X nor Y) chromosomes. In this case, the Y chromosome plays no role in determining sex, although it is involved in the production of sperm in older males. If an individual has two X chromosomes, many female-determining genes result in it becoming a female; if there is only one X chromosome, the relative proportion of male-determining genes is higher, and the individual becomes a male. This arrangement means that males can have both an X and a Y chromosome or just an X chromosome. Grasshoppers and their relatives in the family Orthoptera typically, but not always, have only a single sex chromosome; individuals having two copies (XX) are female, while those having only one (X0) are male.

What about ZZ-ZW individuals? These are found in all birds and some fish, amphibians, and reptiles. In birds, for which we have the most information, females are always the heterogametic sex, possessing one Z and one W chromosome. Unfortunately, the underlying physiological mechanisms for sex determination are unknown, unlike mammals discussed below. One thing that we can say for sure is that the sex chromosomes of birds are not related to those of mammals and evolved separately. Like mammals, the gonads begin life able to become either ovary or testis. Very early on, genes from the Z and W chromosomes are activated, and resulting hormonal changes (probably) direct the development of either sex.

Fish have a multitude of genetic and environmental sex-determining mechanisms.[7] While some have distinct sex chromosomes (e.g. threespine sticklebacks; *Gasterosteus aculeatus*), others are better classified as having 'polygenic' systems where genes on more than one chromosome are involved in determining sexual development. For example, a single locus on three different chromosomes controls sex in the platyfish (*Xiphophorus maculatus*) so that females are XX, XW, or WY, and males are XY or YY. Many African cichlids (e.g. *Labeotropheus spp.*) have a polygenic system involving four chromosomes. In this case, females can be XXZZ, XXZW, or XYZW, while males are XYZZ. And the list goes on.

Certain regions on the sex chromosomes seem to be particularly important in controlling sexual development. In mammals, the two sexes are indistinguishable during the early stages of development. Known as 'indifferent' zygotes, individuals destined to become males or females begin to develop a 'female-like' reproductive system. It isn't until the SRY region of the Y chromosome is activated that we see differences in the development of the sexes. This gene codes for a protein that regulates the production of other proteins necessary for the development of the testes. Note that the concept of a default sex in mammals isn't entirely true. We often hear that embryos will develop as females if no Y chromosome is present. In humans, at least, while rudimentary ovaries may develop in the absence of the Y chromosome, they require a second

X chromosome to become functional. In other words, individuals having only one X chromosome are phenotypically female (XO: Turner's syndrome) but are sterile, and ovaries are nonfunctional.

Few master sex-determining genes, such as the SRY gene, have been discovered in other species. In Japanese medaka, a well-studied fish with an XX/XY sex-determining system, there is a similar region on the Y chromosome called DMY. Birds lack the SRY gene responsible for the development of male characteristics in mammals, and no one has yet uncovered a gene for 'maleness' or 'femaleness' on either the Z or W chromosome. Two hypotheses have been proposed. The first theory is that sex is dose-dependent; an individual possessing two Z chromosomes becomes a male as the number of male-determining genes on the Z chromosome is higher. The second is that a yet-unidentified gene (or genes) on the W chromosome is dominant and results in female traits developing.

7.2.2 Haplodiploidy

The hymenopterans are an insect order that includes ants, bees, wasps, lesser-known thrips, and sawflies, and are second only to the Coleopterans, or beetles, with almost 150,000 described species. Many are best known for their bites and stings (as I discovered first-hand when harassed by ants protecting an acacia bush in Costa Rica). Still, all have a unique genetic method of determining the sex of their offspring, known as haplodiploidy (Figure 7.3).

While this is also a genetic sex-determining system, the sex of individuals is really controlled by their ploidy level. Females have two sets of chromosomes, one set from

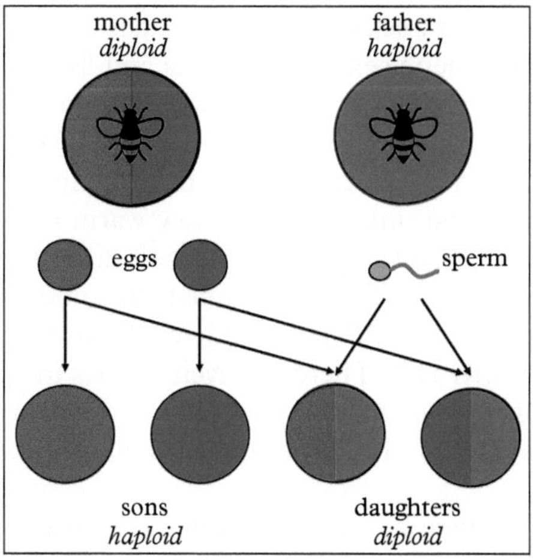

Figure 7.3 *Haplodiploidy is a sex-determining genetic system best known in the Hymenoptera, an insect order that includes ants, bees, and wasps. Unfertilized eggs develop as haploid males, while fertilized eggs develop as diploid females.*

each of their parents, and are, therefore, diploid. On the other hand, males come from unfertilized eggs and only have one set of chromosomes (haploid), all coming from their mothers. It turns out that a single gene determines sex, and it has different alleles. If the individual gets two copies of this gene, it becomes a female; if it gets only one copy, it becomes a male. This method of controlling the offspring's sex would seem like a good system in which mothers could exert control over the sex ratio of their progeny. Being haplodiploid means that mothers are equally related to their sons and daughters; their daughters, who raise the brood, are related to each other by 75%. The different degrees of genetic relationship found within the colony have been used to explain eusociality in many species in this group and potential mother-daughter conflicts over the ideal sex ratio of the brood.

7.3 Environmental sex determination

In many animals, the development of sex is unrelated to their genetic makeup. Instead, the environmental conditions in which they find themselves early in development will determine whether they become male or female. In particular, three main factors affect their sexual development—temperature, location, and density.

7.3.1 Temperature

The patterns of sex determination in reptiles have been extensively studied because they exhibit extraordinary variability among vertebrates. For example, several turtles, some lizards, and all snakes display genetically based sex determination. In these cases, males are sometimes the heterogametic sex (some turtles and lizards, XY of an XY-XX system) and sometimes the homogametic sex (some turtles and lizards and all snakes, ZZ of a ZZ-ZW system). However, reptiles have been primarily studied because many species respond to temperature during the developmental period. Instead of sex being controlled by genetic factors working independently of the environment, the sex of many reptiles is determined after fertilization, influenced by how warm or cold it is. In these cases, the sex chromosomes (if we can call them that) don't differ clearly between males and females, at least in those species closely examined. In some cases, females develop in colder temperatures (most turtles; Pattern 1A), while in other cases, it is males that differentiate (tuatara; Pattern 1B). There are even species in which females develop at high and low temperatures, while males mature at intermediate ones (crocodiles, some geckos, some turtles; Pattern 2). To illustrate, let's look at species indicative of each of these (Figure 7.4).

Hermann's tortoises (*Testudo hermanni*) are often kept as pets, although I expect they aren't very cuddly. Further, they are now listed as a threatened species because of habitat destruction and the removal from their natural surroundings for the pet trade. Native to the southern regions of Europe, a Hermann's tortoise female lays between two and twelve eggs, depositing them in the soil at a depth of about 10 cm. If the eggs develop

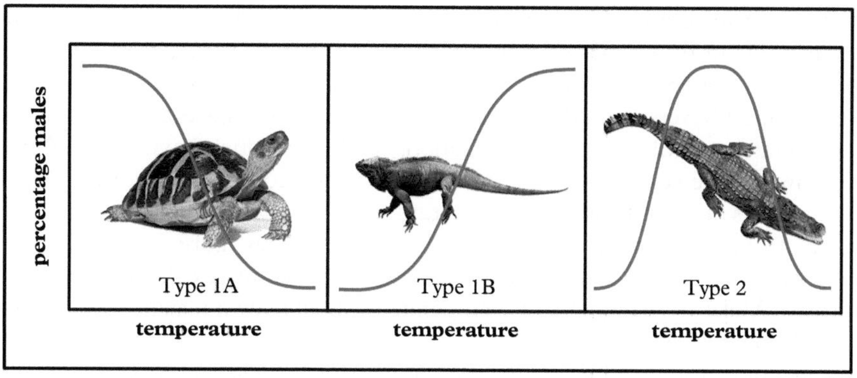

Figure 7.4 *The sex of many reptiles is determined by the temperature at which they develop. In Type 1 species, such as Hermann's tortoise, more males will develop when the temperature is cooler. Type 1b species, like the tuatara, exhibit the opposite pattern. Type 2 species are thought to be ancestral to these two. Female alligators and crocodiles, for example, are more likely to develop at high and low temperatures, while males develop at intermediate temperatures.*

Credits: Studio Empreinte/Adobe Photo ID 311,030,410 (tortoise); JackF/Adobe Photo ID 247,841,290 (tuatara and alligator).

between 25°C and 30°C, all males hatch; at warmer temperatures of 33°C and above, only females will develop.[8] At intermediate temperatures, a mixture of both sexes will emerge from eggs.

Two species of tuatara (*Sphenodon spp.*) demonstrate the rare opposite pattern. Tuataras split from the other modern-day reptiles about 250 million years ago and, although they resemble lizards, are a distinct evolutionary branch. Individuals take around ten years to reach sexual maturity and can live till they are over 100. Reproduction takes a long time, and females can lay about half a dozen eggs every four years, which incubate for about 15 months before they hatch. If eggs of either species are kept at less than 21°C during the critical development period, they are more likely to become females; above 22°C, they become males.[9]

American alligators (*Alligator mississippiensis*) are one of the few reptiles that protect their eggs and offspring. For this reason, it's probably best not to startle a female alligator when her eggs are nearby. After building a nest of vegetation, sticks, and mud, the female lays 30–50 eggs. After depositing her eggs, the mother covers them with more leaves, which give off heat as they decay. She will then watch over the eggs from predators for approximately 65 days until they hatch. Various species of crocodile, of which the American alligator is one, have been studied extensively for their method of sex determination. At first, it was thought that females develop at cooler temperatures and the percentage of developing as males increases as it gets warmer. However, females also hatch more frequently at the highest temperatures, at least for American alligators and all species for which sufficient data exist.[10] Interestingly, since females can adjust the temperature at which the eggs develop, they can potentially influence the sex ratio of the nestlings that hatch (although whether they actually do is uncertain).

In addition to reptiles, temperature-dependent sex regulation was initially thought to occur in about 60 species of fish, although recent evidence suggests that genotype by temperature interactions may control sex determination in many of these species.[11] For example, different flounder species (*Paralichthys spp.*) have a mixture of genetic and temperature-regulated sex determination. While individuals having an XY genotype become males, the sex of fish that are XX is determined by temperature. Fish sexual development is difficult to study under natural conditions, and most of our information has come from laboratory studies. Regardless, there are far more fish species than any other vertebrate, and no doubt more cases of temperature sex determination will be discovered. Males develop at higher temperatures in all known instances of TSD in fish (Type 1B; Figure 7.4).

Let me go off-topic and finish this section with a cautionary note. Climate change is resulting in increased temperatures in many regions, including those in which we find species that rely on temperature-dependent sex determination. Scientists have warned for years that this will almost certainly affect the sex ratios of these species and will quite likely result in population declines and even extinctions.[12] Indeed, this already appears to be happening. In green turtles, females are more likely to hatch at higher temperatures. In nesting grounds in Australia, moderate female biases occur in the southern cooler beaches (65%–69% of hatchlings were female), but extreme biases in hatchlings were found in the warmer, more northern nesting grounds (99.1% of juveniles).[13] The spotted skink (*Carinascincus ocellatus*) is unusual because sex is genetically determined in highland populations but temperature-dependent in the lowlands. Not surprisingly, higher temperatures experienced by lowland populations have resulted in more females; there was no effect in the highlands.[14] Accordingly, the researchers warn that climate change can have unforeseen impacts on species. Sex determination in many species is yet another reason to keep temperature changes associated with climate change minimal.

7.3.2 Location

The green spoon worm (*Bonellia viridis*) (Figure 7.5) is a marine species that occurs off the coast of Europe, where it inhabits the spaces between rocks on the ocean bottom. One of the most distinguishing things about this worm is the remarkable degree of sexual dimorphism. While the female's body is only about 10 cm long, it has a proboscis that can extend for more than a metre to search for microorganisms crawling around the seabed. The females also produce a substance in their skin called bonellin, which, besides making it very green, killing prey, and protecting her from predators, has a masculinizing role on undifferentiated larvae. Males, on the other hand, are tiny. They live symbiotically close to the female's reproductive organs and are only 3 mm long. Their small size allows up to 85 males to live on the proboscis of the female, although only about four will get to live inside her and fertilize the eggs. After fertilization (by how many males isn't clear), the female produces about 1000 eggs that hatch into free-swimming larvae, eventually settling on the substrate. Now it gets interesting. It appears that the sex of some of the larvae is genetically determined, while the sex of most is determined by their settling

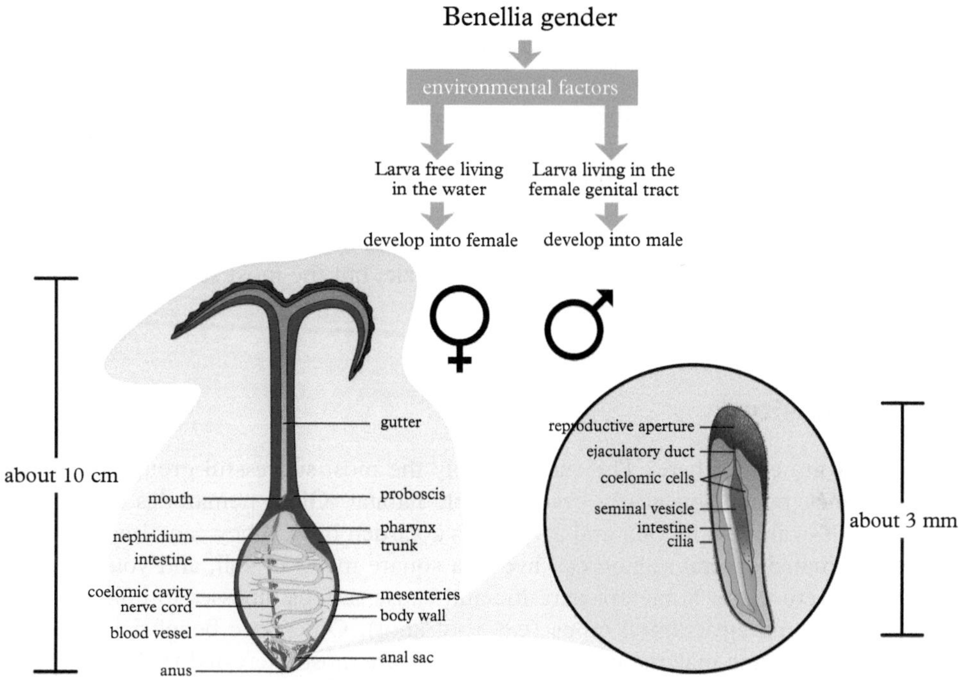

Figure 7.5 *The biological sex of Bonellia individuals is controlled by the location at which they settle. Those that settle in regions touching the proboscis of the female develop as small males and complete their life cycles inside the female. Those that land elsewhere become the much larger females.*
Credit: udaix/Adobe Photo ID 500,760,386.

patterns.[15] If the larva is exposed to bonellin by coming in contact with the female, hormonal events will ensure it becomes a male. If not, it becomes a female. Therefore, if there are very few females around, the larvae are more likely to become females; lots of females mean that larvae are more likely to become males.

Bone-eating worms (*Osedax spp.*), so named because females consume the bones of dead whales, are a group of polychaete worms only discovered in 2002. In most of the 12 species, males are much smaller than females, and the method of sex determination is similar to that of *Bonellia*. For example, females of the species *O. roseus*, attached to dead whale bones, mature quickly, and those harbouring males release fertilized eggs.[16] After hatching, the sexless larvae drift around the ocean in a kind of stasis, often giving them their other commonly used name, 'zombie worms', settling on the same or another whale skeleton. If they land on a whale bone, they become females; if larvae land on a female worm, they become males and enter the gelatinous tubes surrounding females. Each female averages three to four microscopic males that produce enough sperm to fertilize her eggs.

Crepidula fornicata, common marine gastropods that live off the east coast of North America, go by various names, including slipper shells, slipper limpets, and my favourite, fornicating slipper snails. The larva of the slipper shell always starts life as a male. However, if a larva lands on the substrate, it quickly changes from male to female.[17] This female subsequently releases chemical signals attracting more larvae. If they find themselves next to the female, they remain male. Indeed, their sexual development is influenced by masculating pheromones released by the female. As settling by larvae proceeds, mounds of snails are formed, but the oldest and bottommost individual is always a female. Any of these males can fertilize the female, but the most successful male will be the largest that settles next to her.

7.3.3 Density

Nematodes are everywhere. They are arguably the most successful group of animals on the planet, occupying nearly every liveable habitat. Giant nematodes that live in sperm whales can be 8 m long and as thick as a garden hose. They can also be so tiny that an estimated several million can live in a square metre of soil, and you wouldn't know they were there. Some are parasitic and cause serious illnesses in humans (e.g. elephantiasis) and agricultural crops (e.g. root knot). Others are beneficial and contribute to the organic material in soil. But here, we are most interested in them because the individuals of a particular parasitic group use density to determine which sex its members become. The mermithid nematode, *Romanomermis culicivorax*, is a parasite of a specific mosquito species. It and several other parasitic mermithids seem to respond to the population burden of the parasite species already inside the host. More males develop at high population densities, while females are produced at low densities.[18] And to make matters confusing, the sex of the developing young is also influenced by temperature. At temperatures between 20° and 30°C, equal numbers of males and females develop (depending on density), while more females are produced at lower temperatures.

7.3.4 Under what conditions should the environment determine sex?

At the beginning of this chapter, the question was posed whether parents could adaptively manipulate the sex ratio of their offspring. The focus of sex-ratio manipulation has been birds and mammals, two groups where sex is determined genetically. Once the sex of these offspring is set, the external environment plays no role in influencing whether they'll develop as males or females. Not so in species in which genes don't directly determine sex. In these groups, parents might adaptively manipulate the sex of their offspring by allowing them to develop in conditions that favour one sex or the other. But when do certain conditions favour male offspring while others favour female offspring?

Suppose the environment is variable, and the different sexes perform better (i.e. are fitter) in different habitats. In that case, a mother can enhance her fitness by allowing the environment to determine the sex ratio of her progeny; this is known as the Charnov-Bull model.[19] For example, let's use the example of the parasitic mermithids. Population density is often associated with food availability—low density usually means more food. As female mermithids generally require more food to complete their development, it makes sense to develop as a female when food is abundant, as expected when there are fewer individuals.

Let's look at another example. There are many kinds of amphipods, with almost 10,000 species described. Amphipods are small crustaceans that can live in marine or freshwater or even inhabit terrestrial environments as long as those environments are relatively moist. One amphipod has received a great deal of attention because of its unique pattern of sex determination. Known as *Gammarus duebeni* (unlike slipper shells, they have no common name), the sex of developing young seems to be influenced by two different environmental factors—photoperiod and temperature. Males are produced when the photoperiod is short, and the temperature is cooler, as expected at the beginning of the breeding season. This way of determining sex gives males more time to grow, mainly if the breeding season is relatively short.[20] Size affects male fitness more than that of females as large males can better compete for and copulate with many females; large size has no such advantage in females. In other words, males benefit by developing at low temperatures and shorter day lengths. Oddly, in more southern populations, temperature does not affect sex; photoperiod alone determines whether the individual becomes male or female.

It is also possible that different temperatures affect the fitness of males and females differently. While some evidence supports this,[21] much work is needed in this area. Temperature may be significant in determining many phenotypic characteristics and the relative success of individuals when they reach breeding condition. Unfortunately, there is little information on how temperature influences the reproductive success of many reptiles. Many of them are relatively long-lived; long-term observations that link an individual's developmental temperature with their breeding success are difficult to obtain but required.

7.4 Hermaphrodites

Hermes and Aphrodite were the Greek gods of male and female sexuality. Their son, Hermaphrodites, was very handsome and soon caught the eye of a water spirit goddess who prayed to be with him for all time. Her prayers were answered by the gods, who merged the bodies of the two young lovers into one. Since then, the term 'hermaphrodite' has been used to portray an individual with both male and female characteristics. Individuals having ovaries and testes at the same time (simultaneous hermaphroditism) is only one type of hermaphroditism. In some species, males can change to females, or females can change to males (sequential hermaphroditism). Still, fewer than 1% of vertebrates are hermaphroditic, and this trait is primarily confined to fish.[22]

7.4.1 Simultaneous hermaphrodites

Everybody knows what a snail is. There are terrestrial, freshwater, and saltwater snails. In some of these snails, we can find male and female individuals; in other words, the sexes are separate. In others, however, the reproductive organs of both sexes are located in the same individual. You'd think this would be a real advantage—individuals who couldn't find a suitable mate could fertilize themselves. However, snails rarely do. Individuals in these hermaphroditic species, although capable of producing both sperm and eggs, typically find another individual who fertilizes their eggs but also receives sperm to become pregnant themselves. Many of the land species even possess a structure called a 'love dart' that improves their chances of fertilizing the other individual.

Common earthworms (*Lumbricus terrestris*), also known as nightcrawlers or dew worms, are also hermaphrodites, having both male and female reproductive organs. Like many snails, however, they seek out another individual with whom to have sex (Figure 7.6). The male and female gonads of earthworms are located in the head region (if

Figure 7.6 *Earthworm sex can be pretty efficient because each individual has both male and female reproductive organs. However, they don't fertilize themselves, presumably because of the high cost of selfing. Instead, they transfer sperm to another individual while closely connected to them at the clitellum—first one and then the other. Both individuals then produce capsules, which protect the fertilized eggs until they hatch.*

Credit: Mikelane45/Dreamstime Photo ID 33,761,384.

earthworms can be said to have a head), situated so that the individual cannot fertilize itself. Upon meeting another earthworm that wants to reproduce, individuals line up and transfer sperm to their partner. After disconnecting, they each leave behind a small capsule containing fertilized eggs.

Among vertebrates, simultaneous hermaphrodites are relatively uncommon, although not unheard of. Like many of its close relatives, the black hamlet (*Hypoplectrus nigricans*) is a fish with both male and female reproductive organs. Unlike most snails and earthworms, fertilization happens externally in this species, and eggs are planktonic. Individuals spawn in the late afternoon or early evening, a few hours before sunset.[23] Generally, breeding partners alternatively release egg parcels and sperm. This is one of the few groups where individuals can produce eggs and sperm concurrently (and thus are simultaneous hermaphrodites), but reciprocal matings usually occur sequentially.

Finally, I must mention mangrove killifish (*Kryptolebias marmoratus*) as this is one of the few vertebrate species that can (and does) mate with itself. Individuals in most of the population are hermaphroditic, although anywhere from 2%–25% are males.[24] In this respect, it represents one of the few androdioecious species (having both hermaphrodites and males in a population). There are no purely female individuals, and hermaphrodites have never been observed to mate with other hermaphrodites; hermaphrodites can either fertilize themselves (as they often do) or can be fertilized by a male. So even though another hermaphrodite has the male bits necessary to fertilize them, they don't seem to be that attractive.

7.4.2 Sequential hermaphrodites

'Finding Nemo' was a great film based on many species that really exist. I especially liked the scene where the anglerfish lures Nemo and Dory to the depths with its glowing lure (a modified dorsal ray on the head). However, one thing that wasn't portrayed very accurately was the mating system that gave rise to Nemo in the first place. Anemone fish, or clownfish, are protandrous hermaphrodites. All fish begin life as males and may become females should the need arise. Groups of clownfish, consisting of many males and one large female, live in the protective tentacles of an anemone. Even though many males are around, the female will only mate with one. If she dies, the breeding male changes sex and becomes a female, and the largest of the other males then becomes the breeding male.

Protogynous species are just the opposite of clownfish. Instead of becoming female at some point in life, individuals start as females and then change to males if needed. Such a system is much more common in fish, probably because a male's reproductive success increases with age while that of females doesn't or increases only minimally. In other words, if you are going to be an old individual, it pays to be a male (evolutionarily speaking, that is). Probably for this reason, about 66% of the 461 species known to be hermaphroditic in fish are protogynous, while only 12% are protandrous.[22]

Bluehead wrasses (*Thalassoma bifasciatum*) are abundant cleaner fish that remove the dead skin and parasites from often much larger fish that come to their 'cleaning stations'. Both species benefit from this interaction—the 'client' gets cleaned, and the wrasse receives a free meal. But these wrasses are probably best known for their ability to change sexes, from a female to a male. Individuals start life as either a male or a female, but most dominant females can change sex. This change often occurs in response to the death or disappearance of the dominant male, the most successful breeder.

Usually, once a sex change has occurred, there is no going back. A clownfish that has gone from being a male to a female cannot become a male again. Similarly, a male bluehead wrasse, formerly a female, cannot return to its previous life. Sex change usually takes you in one direction and happens only once. However, there are a few species that buck this trend. A coral goby is one of several species of coral-dwelling fish that form monogamous pairs. Individuals can change sexes in these species depending on their potential breeding partner. It has been suggested that this may be due to the high cost (energetically and because of predation) of moving between corals.[25] In other words, you want to ensure that you only have to move once when it comes time to breed and not be restricted by the sex of the fish that already reside there.

7.4.3 When should individuals change sex?

Why do individuals change sex? I've mentioned one reason—known as the size advantage model. This model suggests that the reproductive success of a larger or older individual is disproportionately advantageous to either males or females. In other words, when you are old and big, you are better off being the sex that benefits most by being older or larger. This size advantage would explain why protogyny is favoured when species display harem polygyny, as bigger males are expected to mate with more females. Alternatively, the low-density model predicts that if you belong to a population in which individuals have dispersed widely, it would be helpful to be able to mate with the first individual encountered. However, while this ability might be beneficial to simultaneous hermaphrodites, a sequential hermaphrodite probably can't change sex fast enough for it to matter. While there is some support for both models, there are also exceptions, and it's likely that the evolutionary reason behind the ability to change sex depends on the ecology of the species.

Changing one's sex, at least if you are one of the hermaphroditic fish species, is not unusual. But going from a male to a female, or vice versa, involves a significant transformation. Not only does it require converting the gonads, but numerous other behavioural, anatomical, neural, and hormonal changes must occur. The perception of environmental cues first triggers a sex conversion, but a cascade of neuroendocrinal changes, likely involving the pituitary-gonadal axis, is responsible for completing the process. Regulation of the various molecules instrumental in this pathway is vital in determining the sex of certain animals during sexual development and in converting from one sex to another. As it turns out, increasing or decreasing the activity along this pathway can result in sex changes in older individuals. For example, if an aromatase inhibiter is used to reduce the estrogen produced in honeycomb groupers (*Epinephelus merra*), a type of protogynous fish, fully functional males will develop from females.[26] Importantly, by regulating estrogen levels, sex change in both directions can be controlled. There are undoubtedly significant costs to changing sex, and the molecular pathways that lead to sex change have not been well worked out.

7.5 Summary

There are many differences between males and females, but how the sexes develop in different groups is variable. In some cases, the sex of an individual is determined by chromosomes. Sex chromosomes are generally referred to as **X** and **Y** or **W** and **Z**, depending on which sex is heterogametic. In some groups, such as the ant, wasps and their relatives, females have two sets of chromosomes, while males have only one. However, the biological sex of all individuals controlled genetically is fixed at fertilization, and sex-specific anatomical, physiological, or neurological differences arise during development. In other groups, the sex of individuals is environmentally determined. Usually, the embryo's temperature during a sensitive period during growth determines its sex; this is the case for many reptiles and some fish. However, the density and location of individuals may also play a role. In these cases, the sex of the individual is determined at some point after fertilization occurs.

Although male and female reproductive organs are usually found in different individuals, they may be found in the same individuals in certain species. In some, individuals may even change sex during their lifetime, going from male to female or female to male when conditions are more likely to favour being a specific sex. Whether sex is determined genetically, environmentally, or can change over time, there are often situations in which the sexes have different fitness; some hypotheses are presented that may explain observed patterns of sex allocation and determination.

References

1. Fisher RA. The genetical theory of natural selection. 2nd ed. New York: Dover Publications; 1958, p. 291.
2. Trivers RL, Willard DE. Natural selection of parental ability to vary the sex ratio of offspring. Science. 1973 Jan;179:90–92.
3. Valterová R, Procházka P, Požgayová M, Piálková R, Piálek L, Šulc M, et al. Son or daughter, it does not matter: Brood parasites do not adjust offspring sex based on their own or host quality. J Ornithol [Internet]. 2020 May [cited 2023 Jan 30];161:977–986. Available from: https://doi.org/10.1007/s10336-020-01782-9
4. Douhard M. Offspring sex ratio in mammals and the Trivers-Willard hypothesis: In pursuit of unambiguous evidence. BioEssays [Internet]. 2017 Jul [cited 2023 Jan 16];39(9): 10.1002/bies.201700043. Available from: https://doi.org/10.1002/bies.201700043
5. Katsuki M, Harano T, Miyatake T, Okada K, Hosken DJ. Intralocus sexual conflict and offspring sex ratio. Ecol Lett [Internet]. 2012 Jan [cited 2023 Mar 3];15(3):193–197. Available from: https://doi.org/10.1111/j.1461-0248.2011.01725.x

6. Leonard JL, editor. Transitions between sexual systems [Internet]. Switzerland: Springer; 2018 [cited 2023 Mar 3], p. 363. Available from: https://link.springer.com/book/10.1007/978-3-319-94139-4
7. Godwin J, Roberts R. Environmental and genetic sex determining mechanisms in fishes. In: Leonard JL, editor. Transitions between sexual systems [Internet]. Switzerland: Springer; 2018 [cited 2023 Mar 3]. p. 311–344. Available from: https://link.springer.com/book/10.1007/978-3-319-94139-4
8. Eendbak BT. Incubation period and sex ratio of Hermann's tortoise, Testudo *hermanni boettgeri*. Chelonian Cons Biol [Internet]. 1995 [cited 2023 Jan 24];1(3)227–231. Available from: https://chelonian.org/wp-content/uploads/file/CCB Vol 1 No 3 (1995)/Eendebak_1995.pdf
9. Mitchell NJ, Nelson NJ, Cree A, Pledger S, Keall SN, Daugherty CH. Support for a rare pattern of temperature-dependent sex determination in archaic reptiles: Evidence from two species of tuatara (Sphenodon). Front Zool [Internet]. 2006 Jun [cited 2023 Jan 31];3:9. Available from: https://doi.org/10.1186/1742-9994-3-9
10. González EJ, Martínez-López M, Morales-Garduza MA, García-Morales R, Charruau P, Gallardo-Cruz JA. The sex-determination pattern in crocodilians: A systematic review of three decades of research. J Anim Ecol [Internet]. 2019 Sep [cited 2023 Jan 30];88(9):1417–1427. Available from: https://doi.org/10.1111/1365-2656.13037
11. Shen ZG, Wang HP, Yao H, O'Bryant P, Rapp D, Zhu KQ. Sex determination in bluegill sunfish *Lepomis macrochirus*: Effect of temperature on sex ratio of four geographic strains. Biol Bull [Internet]. 2016 Jun [cited 2023 Jan 13];230(3):197–208. Available from: http://www.jstor.org/stable/24878832
12. Lockley EC, Eizaguirre C. Effects of global warming on species with temperature-dependent sex determination: Bridging the gap between empirical research and management. Evol Appl [Internet]. 2021 Oct [cited 2023 Jan 25].14(10):2361–2377. Available from: https://doi-org.proxy.library.upei.ca/10.1111/eva.13226
13. Jensen MP, Allen CD, Eguchi T, Bell IP, LaCasella EL, Hilton WA, et al. Environmental warming and feminization of one of the largest sea turtle populations in the world. Curr Biol [Internet]. 2018 Jan [cited 2023 Jan 24];28:154–159. Available from: https://doi.org/10.1016/j.cub.2017.11.057
14. Cunningham GD, While GM, Wapstra E. Climate and sex ratio variation in a viviparous lizard. Biol Lett [Internet]. 2017 May [cited 2023 Jan 24];13:20170218. Available from: http://doi.org/10.1098/rsbl.2017.0218
15. Berec L, Schembri PJ, Boukal DS. Sex determination in *Bonellia viridis* (Echiura: Bonelliidae): Population dynamics and evolution. Oikos [Internet]. 2005 Mar [cited 2023 Jan 25];108(3):473–484. Available from: http://www.jstor.org/stable/3548792
16. Rouse GW, Worsaae K, S. Johnson SB, Jones WJ, Vrijenhoek RC. Acquisition of dwarf male 'harems' by recently settled females of *Osedax roseus* n. sp. (Siboglinidae; Annelida). Biol Bull [Internet]. 2008 Feb [cited 2023 Jan 25];214(1):67–82. Available from: https://www.jstor.org/stable/25066661
17. Proestou DA, Goldsmith MR, Twombly S. Patterns of male reproductive success in *Crepidula fornicata* provide new insight for sex allocation and optimal sex change. Biol Bull [Internet]. 2008 Apr [cited 2023 Jan 25];214(2):194–202. Available from: https://doi.org/10.2307/25066676

18. Tingley G, Anderson R. Environmental sex determination and density-dependent population regulation in the entomogenous nematode *Romanomermis culicivorax*. Parasitology [Internet]. 1986 Apr [cited 2023 Jan 25];92(2):431–449. Available from: https://doi.org/10.1017/S0031182000064192
19. Charnov E, Bull JJ. When is sex environmentally determined? Nature. 1977 Apr;266:828–830.
20. Dunn AM, Hogg JC, Kelly A, Hatcher MJ. Two cues for sex determination in *Gammarus duebeni*: Adaptive variation in environmental sex determination? Limnol Oceanogr [Internet]. 2005 Jan [cited 2023 30 Jan];50(1):346–353. Available from: https://doi.org/10.4319/lo.2005.50.1.0346
21. Steele AL, Warner DA. Sex-specific effects of developmental temperature on morphology, growth and survival of offspring in a lizard with temperature-dependent sex determination. Biol J Linn Soc Lond [Internet]. 2020 Jun [cited 2023 Jan 30];130(2):320–335. Available from: https://doi.org/10.1093/biolinnean/blaa038
22. Kuwamura T, Sunobe T, Sakai Y, Kadoa T, Sawada K. Hermaphroditism in fishes: An annotated list of species, phylogeny, and mating system. Ichthyol Res [Internet]. 2020 May [cited 2023 26 Jan];67:341–360. Available from: https://doi.org/10.1007/s10228-020-00754-6
23. Fischer EA. Mating behavior in the black hamlet—gamete trading or egg trading? Environ Biol Fish [Internet]. 1987 Feb [cited 2023 Jan 26];18:143–148. Available from: https://doi.org/10.1007/BF00002602
24. Mackiewicz M, Tatarenkov A, Turner BJ, Avise JC. A mixed-mating strategy in a hermaphroditic vertebrate. Proc Biol Sci [Internet]. 2006 Oct [cited 2023 Jan 26];273:2449–2452. Available from: https://doi.org/10.1098/rspb.2006.3594
25. Sunobe T, Sado T, Hagiwara K, Manabe H, Suzuki T, Kobayashi Y, et al. Evolution of bidirectional sex change and gonochorism in fishes of the gobiid genera *Trimma, Priolepis*, and *Trimmatom*. Naturwissenschaften [Internet]. 2017 Mar [cited 2023 Jan 26];104(3–4):15. Available from: https://doi.org/10.1007/s00114-017-1434-z
26. Bhandari RK, Higa M, Nakamura S, Nakamura M. Aromatase inhibitor induces complete sex change in the protogynous honeycomb grouper (*Epinephelus merra*). Mol Reprod Dev [Internet]. 2004 Mar [cited 2023 Jan 26];67(3):303–307. Available from: https://doi.org/10.1002/mrd.20027

8

Conflict Between the Sexes

Some male desert spiders (*Stegodyphus lineatus*; Figure 8.1) mature later than females, only to find a potential breeding partner looking after one hundred or so eggs that another male had fertilized.[1] The females of this species go out of their way to care for the one brood they produce, feeding and protecting their offspring for two weeks after they hatch. The young express their gratitude by eating their mother. Although females only produce a single brood during their lives, they can generate replacement clutches if their eggs are destroyed. Given the context, this is what the late-maturing males may do. Such a male doesn't come across a female without eggs very often, so to ensure that he gets to copulate, he may fight with a previously mated female and destroy her clutch of eggs. In fact, males may be responsible for one-third of all egg losses in this species. However, by eliminating her current eggs, he gets to mate and pass on his genes. Unfortunately, she pays a high price as the number of young she produces decreases with every clutch. Obviously, the interests of the males and females of this species conflict. But is this type of antagonism between the sexes commonly observed?

8.1 What's good for one sex . . .

When it comes to sexual reproduction, males and females face a dilemma. On the one hand, they need to co-operate to a certain extent if they are going to combine their genes to produce offspring. When fertilization is external, both sexes need to release their gametes around the same time and place for reproduction to be successful. If fertilization is internal, males and females may have to co-operate to the extent that sperm is introduced into the female reproductive tract (although the method of doing this may not always be mutually beneficial). If further care by both parents is necessary for the offspring's well-being, adult males and females must co-operate, at least to a degree, to help rear their young. So, producing offspring when biparental care is required has historically been considered a co-operative venture, where males need females and females need males; only by working together can they pass on their genes to the next generation. I will discuss biparental care of the young in more detail in

The Evolution of Sex. Kevin Lee Teather, Oxford University Press. © Kevin Lee Teather (2024).
DOI: 10.1093/9780191994418.003.0008

Figure 8.1 *Male desert spiders will often destroy the clutch of previously mated females so that they can fertilize all of the new eggs they produce. Although females typically produce only one clutch of eggs during their lives, they will lay replacement ones if their first clutch is lost. However, the number of eggs they have declines, and their fitness suffers because of this destructive behaviour by males. In this, and many other cases, methods to achieve optimal fitness differ for males and females.*
Credit: Sarefo (assumed)/CC BY-SA 3.0.

the next chapter as, indeed, it does require co-operation. However, it also can involve conflict.

We have seen previously that differences between males and females are first defined by gamete size. Females invest a substantial amount of energy into the production of an egg; typically, it needs first to attract the sperm and second to meet the nutritional demands of the early embryo. The male's investment in the sperm, on the other hand, is minimal. It generally requires a method of movement but donates little more than genetic material to the fertilized zygote. This early difference in investment into the shared offspring means that sexual selection already operates differently in males and females. The reproductive potential of females is usually limited by the number of offspring they can produce (I say usually because sometimes female success might be determined by the number of males around). On the other hand, male success tends to be limited by the number of eggs they can fertilize, often found in different females. Given that these are the main differences between males and females when there is no further parental care, what characteristics should these be selected for in each sex? Females should develop a more discriminating examination of potential male partners. After all, they have to combine their genes contained in a limited number of eggs with another individual, so must ensure their partners have genes of high quality. Males should develop characteristics that help them fertilize more eggs. These include characters that help them compete with other males for access to females themselves or the resources females need to breed successfully. It also means having characteristics that make them more likely to be chosen by females as breeding partners.

Thus far, I've highlighted a few ways concerning how males and females differ in their approach to sexual reproduction. Reproductive conflicts between the sexes can occur before fertilization (prefertilization), around fertilization, and after fertilization (postfertilization). Below, I provide a few examples that illustrate reproduction may not always be a co-operative venture, but males and females can often achieve the greatest success in very different ways. This list is by no means exhaustive; my intention is simply to show that the paths to optimal evolutionary fitness of the sexes may not only differ but can often negatively affect the reproductive performance of the other sex. First though, I want to briefly go over what is happening at the genetic level to provide a better understanding of how this sexual conflict might work.

8.2 A Genetic primer on sexual conflict

Wouldn't it be nice if males and females always got along? Unfortunately, theoretical reasons and practical studies suggest that the two sexes are often at odds, behaviourally and morphologically, during both breeding and some non-related breeding activities. This is kind of strange. Genetically, males and females are pretty much the same, differing only in their sex chromosomes (if indeed they have sex chromosomes). Think about it. In sexually reproducing species, half an individual's chromosomes come from the mother and half from the father. Although there is often no way to distinguish the sexes during early development, clear differences in anatomy and behaviour generally emerge between males and females as they approach and reach sexual maturity. These differences are not just apparent in the structure of gonads but potentially, for example, in size, colour, and ornamentation. Given these constraints, what is optimal for one sex will not likely be optimal for the other. For example, males may benefit by having a long tail. Even though it may hinder their ability to fly, it is desirable to females who mate more often with males having one. Females don't need a long tail. Not only does it decrease their flying ability, but it doesn't impress males; males will mate with pretty much anything, long tails or not. One might expect all the genes associated with differences between the sexes to be found on the sex chromosomes. But they're not. Additionally, when males and females share any of the genes responsible for a structure or behaviour with a different optimal expression between each sex, we expect some conflict to occur. We also expect conflict to happen when males and females have different interests when they directly interact. These two possibilities are referred to as intra- and interlocus conflicts.

8.2.1 Intralocus conflict

In some cases, sexual conflict occurs when both males and females express a trait, but the fitness optimum for that trait differs between the sexes. Genetically, an allele of the gene that is best for one sex isn't the best for another. Because the gene that codes for this trait is the same for both sexes, it is referred to as **intralocus conflict** (Figure 8.2). Of course, more than one gene likely codes for the trait, but let's stick to one gene as it's the simplest scenario. This conflict has been described as a 'tug-of-war' between males and females for the genotype's phenotypic expression and often results in sexual dimorphism between the sexes.

A nonreproductive example of this might be helpful. The reed warbler (*Acrocephalus scirpaceus*) is a rather small, plain, buff-coloured bird extensively studied in Europe. It has various dietary preferences, mainly insects, although it has been known to forage on berries and seeds. During the breeding season, the female builds a nest, lays four or five eggs, and both parents feed the hatched offspring. This bird is particularly well-known for being a host of the European cuckoo, which lays a single egg in many of the reed warbler nests and, after hatching, ejects the warbler eggs and nestlings, receiving all the food from the warbler parents. During the nonbreeding season, this warbler undergoes a long-distance migration, so the size of the wings is significant. Males and females are slightly dimorphic when it comes to wing size, indicating that larger wings are better for the male's overall lifestyle while smaller wings may better suit the female. It also suggests that wings may be a target for conflict between the sexes. In other words, males may be

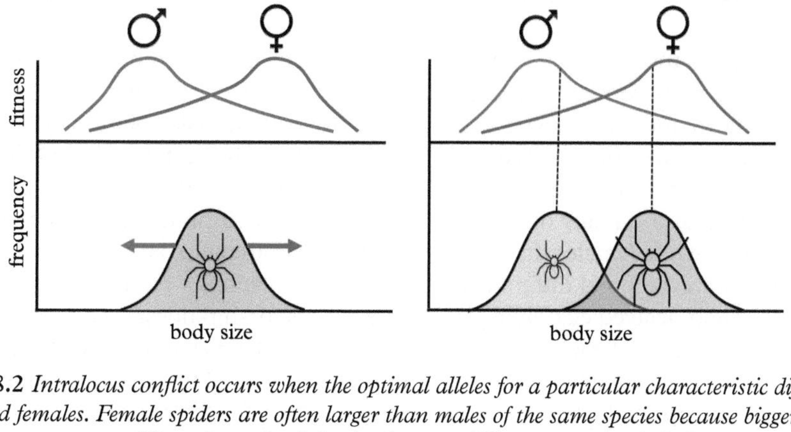

Figure 8.2 *Intralocus conflict occurs when the optimal alleles for a particular characteristic differ in males and females. Female spiders are often larger than males of the same species because bigger females can produce more eggs. Thus, the optimal size for the sexes differs, as seen by their fitness curves at the top. This results in a conflict whereby the best alleles for determining size are different for each of the sexes, creating a 'tug-of-war', and the optimal body size for either sex is not reached.*

better off with even larger wings and/or females with even smaller wings, but inheriting the same genes for wing size restricts the overall differences between the sexes. And this appears to be true. One long-term study found that selection on wing length was different for males and females.[2] Males with longer wings had higher fitness, and females with shorter wings were fitter. Presumably, the gene (or genes) responsible for the length of wings is similar for both sexes. Therefore, the allele(s) determining longer wings is favoured in males while that coding for shorter wings is favoured in females; the optimal wing length may never be reached in either sex. The best allele for a trait will depend in whichever sex it finds itself. In this species, wing size is a trade-off between what's best for males and females.

We often think of natural selection as choosing between traits that give individuals the best chance of success in a particular environment. Thus, individuals are driven towards some optimum value. However, these values may be different for males and females. Intralocus sexual conflict may be an important limiting factor in the evolution of individuals of a species; wings of reed warblers (e.g.) are always going to be shorter than optimal for males, while they are always going to be longer than is best for females. Of course, as we learn about how genes are expressed, this may not be the complete story. Genes may be turned off and on, or their activity modified by other aspects of the genome or individual; if these modifiers are sex-specific, the evolutionary conflict between the sexes may be reduced.

8.2.2 Interlocus conflict

In some cases, the outcome of a direct interaction between a male and a female is beneficial to one and detrimental to the other. It is termed **interlocus conflict** because such interactions often involve different genes in males and females. One or both sexes suffer reduced fitness because of interacting with the other. Thus, coevolution occurs between males and females where a locus for one trait may benefit one sex but is countered (at least partially) by the evolution of another characteristic at a different locus in the other.

Perhaps the most common occurrence of interlocus sexual conflict involves the preferred mating frequency of males and females. In general, males benefit by frequently copulating as they are often limited by the number of eggs they can fertilize. Onion thrips (*Thrips tabaci*; Figure 8.3) are a worldwide agricultural pest that can damage onions, potatoes, tobacco, and many other crops. In this species, virgin females can produce all male offspring, while females that mate sexually have both males and females. However, males and females respond differently to repeated copulations.[3] Females suffer reduced survivorship and delayed egg-laying, while male survivorship doesn't change with mating frequency. Much of the mating cost to females is likely associated with high levels of male harassment, which could occur 20 times per hour. This example is just one of many where the optimal frequency of copulation differs for the two sexes.

Let me give you another example. One of the best-known cases of interlocus sexual conflict involves male and female fruit flies (*Drosophila melanogaster*). Male fruit flies produce a protein that they transfer to the female in their seminal fluid.[4] This isn't a

Figure 8.3 *Male onion thrips like to copulate. Females, not as much. The reproductive success of females declines with repeated copulations, presumably because of the continual harassment by males. The difference in optimal copulation rates is a common form of sexual conflict—the advantages of repeated copulation often differ between males and females.*
Credit: Tomasz Klejdysz/Shutterstock Photo ID 1,416,146,774.

protein that females have, and one of its functions is to elevate her rate of egg-laying. At first, this would appear to benefit both sexes. However, it also has many negative effects on female fitness, the most significant of which is to decrease her receptivity to other males and her lifespan. So, while it increases the male's immediate reproductive success, it reduces the lifetime fitness of females. What can females do? Not much. They need the sperm of males to fertilize their eggs, so they are forced to accept the seminal fluid despite its detrimental effects.

The actual genetics of sexual conflict have not been well-studied, as often multiple genes are involved and tracing them to particular behaviours or structures is challenging. However, numerous examples of conflict exist. Let's look at some of the cases that occur at different points of the reproductive cycle. This list is not exhaustive, but after you read through the examples, it should be clear that males and females have their own reproductive interests at heart, and their behaviour in achieving these does not always coincide with that of the other sex.

8.3 Prefertilization sexual conflict

Much of the conflict that occurs between males and female comes prior to fertilization. This conflict can be witnessed during many of the intersexual interactions during courtship. A few are presented here.

8.3.1 Nuptial gifts

The term 'nuptial gifts' is used here to describe any product given by the male to the female to increase the probability of copulation. Females gain direct fitness benefits if substances provided in the gift can improve the number or quality of their offspring or help them select the best possible mate. Male fitness can increase because mating opportunities are increased. However, male gift-giving provides an opportunity for deception, as demonstrated in the following example.

We've seen that sex can be a dangerous undertaking for many male spiders. A female can mistake her much smaller suiter for prey, often ending poorly for him. This isn't the case in the nursery web spider (*Pisaura mirabilis*), where males and females are similar in size. However, the male still wants to keep the female busy so that he has sufficient time to transfer his sperm. To occupy the female, the male may offer her a wrapped edible gift before sex. Catching (usually flies) and wrapping these gifts can be time-consuming and energy-demanding, and males do what they can to minimize their cost and still mate. In one investigation,[5] males were classified by the type of gifts they provided: (i) no gift, (ii) worthless gift, (iii) regular gift, and (iv) protein-enriched gift. The time of copulation (coinciding with sperm transfer and subsequently mating success) was lowest, not surprisingly, for the males who provided no gifts and longer for gift-giving males regardless of the gift quality. But while females were just as likely to copulate with males providing high- or low-quality gifts, they terminated copulation earlier if the gift was of low quality. Wrapping up the gift seemed to allow the male to deceive the female, albeit temporarily, into mating. Males and females may disagree over copulation and its duration, resulting in strategies and counterstrategies by each sex.

Spermatophores are packages of sperm transferred to the female and used in a wide range of animals, particularly certain species of insects, gastropods, and amphibians. They are also often relatively nutritious, and females can invest the energy they obtain from them into their offspring. In the black field cricket, males provide the female with a spermatophore that is attached directly to her abdomen.[6] In this case, the spermatophore doesn't provide her with nutrition but prostaglandin precursors, which stimulate egg-laying. The duration of spermatophore attachment determines how much sperm is transferred to the reproductive tract of the female, at least to a certain point, and significantly influences the reproductive success of males. Males and females behave differently to the presence of spermatophores. Females can remove spermatophores and select a different mate, using spermatophore acceptance in mate choice. Males will harass the females and attempt to stop them from spermatophore removal. When males are allowed to harass females more (experimentally), there is less opportunity for females to pick the male of their choice. Thus, sexual conflict between the sexes can influence the intensity of sexual selection on female choice.

8.3.2 Copulatory plugs

Mating plugs are formed by the seminal fluid of many males, mainly in certain species of mammals, reptiles, arachnids, and insects. Immediately after mating, the male inserts a substance into the female reproductive tract that forms a plug; this plug partially or wholly prevents another male from fertilizing the eggs, thus increasing the first male's chance of inseminating her. For example, in mice (*Mus musculus*) allowed multiple mates, a copulatory plug inserted by the first male completely prevents fertilization by the second.[7] These plugs are usually gelatinous but may harden after being placed in the female.

Females may not benefit to the same extent as males by repeated copulation, but this isn't to say that they don't benefit at all. In many cases, the reproductive success of a female will increase if she mates more than once. For example, female honeybees (*Apis mellifera*) may mate more than 50 times, although the average is about 14. It pays for a female to mate with several males since increased genetic variability of her offspring results in, you guessed it, lower levels of parasitism.[8] Parasitism by mites and other pathogens has resulted in the decline of this economically important species across North America and is an essential consideration during reproduction. Of course, this wouldn't be good for the individual males with whom she's mated. By mating repeatedly, she decreases the number of offspring sired by any previous mate. Male bumblebees (*Bombus terrestris*) prevent multiple matings by females by inserting a copulatory plug into their reproductive tract, preventing females from remating and increasing the probability that the males' sperm will be successful in fertilization.[9] Therefore, females in this species usually mate just once, although it has been experimentally demonstrated that parasitism rates will decline if they are permitted to copulate with other males.

8.3.3 Forced copulation

Forced copulation is something we don't usually think about in nonhuman animals, but it does happen, and not infrequently. Typically known as FEPCs (forced extrapair copulation), this aggressive tactic used by males to inseminate females occurs in different vertebrates and invertebrates. The theoretical reasons for this behaviour usually revolve around the potential advantages to certain males in fertilizing the eggs of females who are not their mates. Males can reduce the costs of an extended courtship and increase the benefits of siring more offspring. On the other hand, they increase the potential costs of getting injured by the female or her partner. Females presumably do not derive benefits, as is shown by their resistance to forced copulation. They are prevented from assessing the males, they may be seriously injured, and their eggs may be fertilized at an inconvenient time. For example, female fruit flies forced to mate have fewer offspring, higher mortality, and more injuries.[10] Male fruit flies, who often harass females until they copulate, do not benefit greatly but are still rewarded with higher reproductive success. Forced copulation is a clear case of a behaviour being advantageous for certain males but disadvantageous to females.

Copulation by birds is usually a relatively peaceful event. This is primarily due to the similarity in external genitalia exhibited by both males and females in most species. The male clambers onto the back of the female; the two sexes align their cloacae; and the male then transfers the sperm from his reproductive tract to that of the female. The process usually lasts only a few seconds and is often called a 'cloacal kiss'. The important thing is that both sexes must co-operate for copulation to be successful. However, the males of about 3% of the approximately 10,000 bird species have a penis, including the waterfowl (ducks, geese, and swans) and ratites (ostriches and emus). This arrangement means that if females can be held in a suitable position, the males can insert their penis into the female's reproductive tract and transfer sperm without her co-operation.

Figure 8.4 *The ducks are one of the few groups of birds in which males have a penis. They are also among the few groups where males often gang up and force females to copulate. But females make such behaviour difficult by having a corkscrew-shaped vagina, discouraging easy entry. In fact, in species where males have the longest penis, females have the most elaborate vagina. In this photo, two male mallards attempt to force a female to have sex.*

Credit: Pawel Korzekwa/Shutterstock Photo ID 2,285,209,501.

And many do. Indeed, cross-species comparisons of the waterfowl have shown that species with the longest penis are likelier to engage in FEPCs.[11] Mallards are generally socially monogamous, and the only way for a mated male to increase his reproductive output is to have sex with a female other than his partner. Females resist these advances and may suffer injuries, although the seriousness of these injuries is debated.[12] Regardless, female mallards fight back against unwanted males in more ways than one. Anatomically, they have developed a corkscrew-shaped vagina that deters quick access by unwanted males (Figure 8.4).

Female reluctance to mate with a male is also often observed in invertebrates. Water striders (*Gerris gracilicornis)* are a particularly well-studied group when it comes to the examination of the mating conflict between sexes. Most of us have seen water striders. They are those long-legged insects, often called 'pond skaters 'or' Jesus bugs', that can stand and move about on the water's surface. Female water striders want to avoid sex. Perhaps it would be more accurate to say that females wish to avoid having sex repeatedly as they can store the sperm from a single male for weeks and use it to fertilize all their eggs. They have little to gain by mating with additional males, and it may be detrimental to do so. While copulating itself may make females more vulnerable to predators, males of many species will also ride on their backs for minutes to days afterwards, decreasing the female's ability to gather food. So, females have developed an anatomical genital shield that allows them to avoid having sex when they have enough sperm.[13]

Now, let's look at it from the male's perspective. Male water striders can increase their fitness, or the number of offspring they sire, by mating repeatedly. It is also to his advantage to be the last male to copulate with the female as he is most likely to fertilize the eggs, regardless of whether females have stored the sperm from a previous mating. Usually, males simply mount other individuals, males or females, with little or nothing in the way of courtship. If they happen upon an unwilling female, she covers her genital opening with a protective shield, thus thwarting his sexual advances. Frustrated by her unco-operative behaviour, the male creates vibrations along the water surface by tapping his front legs. This behaviour attracts the attention of predators and can be considered a form of extortion; if she fails to copulate, she runs the risk of being eaten.[14] Such strategies and counterstrategies are a hallmark of sexual conflict.

8.3.4 Genitalia

I'm not fixated on genitalia, but they often demonstrate that sex between individuals can harm one of them, usually the female. The male penis is designed to increase the chance of fertilization. Generally, this would also be better for the female; after all, she wants her eggs to be fertilized. However, any injury caused by the penis that makes it less likely that a female will copulate again may lower her lifetime fitness. I've already touched on 'traumatic insemination', where the male bypasses the female's reproductive tract, thrusts his penis into the female's abdomen, and releases his sperm. Male bedbugs are known to do this. However, females don't suffer the extensive costs as

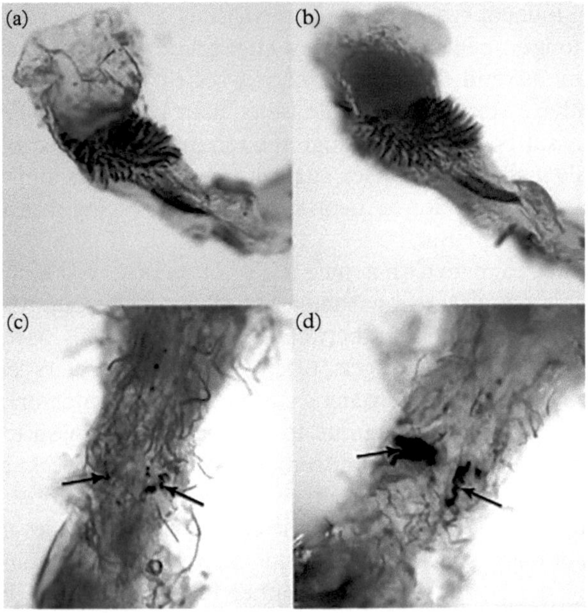

Figure 8.5 *The penis of the male seed beetle is covered by hooks and spines that can damage the female reproductive tract during copulation. This photo shows (a) the penis of a male seed beetle from an individual having spines 30% shorter than those of (b) other males. The damage they inflict on females is (c) less than (d) that inflicted by males with more substantial spines, as demonstrated by the melanized scars (shown by arrows) on the female reproductive tract.*

Credit: Hotzy and Arnquist; permission from Elsevier.

expected. The male generally pierces the abdomen in a specialized region called the spermalege; multiple piercings here did not decrease female fitness.[15] However, any piercing by the male genitalia outside this region resulted in a 50% reduction in egg production.

Regardless, if a female mates repeatedly, she wants her reproductive system to be in tip-top shape. The genitalia of a male seed beetle (*Callosobruchus maculatus*; an aedeagus) is an unpleasant-looking structure closely resembling a medieval torture device (Figure 8.5). It is covered with elaborate spines that help the male grasp the female (internally) while he's transferring his sperm. In fact, those males having the longest and most elaborate spines achieve the highest fertilization rates.[16] You can imagine that this damages the female reproductive tract (the bursa copulatrix). More damage is done the more often she copulates, and the number of eggs she produces declines. Females can minimize the damage by making a very tough bursa. Not surprisingly, females in those species in which males have the most elaborate spines have the toughest and thickest reproductive tract.[17]

Male snakes have hemipenes, a pair of penis-like structures typically inverted inside the body, extended away from the body only during copulation. The hemipenes are characterized by having many hooks and spines, with sperm travelling along a groove outside the structure into the female's reproductive tract. Male red-sided garter snakes have a multispined hemipenis with a sizeable basal spine that is inserted into the female during the reproductive bout. This spine is thought to anchor the intromittent structure in the female cloaca when she attempts to end copulation by rolling her body. When spines are removed experimentally, the duration of copulation is shorter.[18]

8.3.6 Mate guarding

Males often attempt to increase the chance of fertilizing a female by preventing other males from copulating with her. Often, this means remaining attached to the female before copulation or after copulation is complete. For example, in the harlequin toad (*Atelopus laetissimus*), males may remain attached to females for 70–135 days.[19] The necessity of mate guarding relates to the degree of competition with other males who often attempt to dislodge him. Indeed, it isn't unusual for males to clasp females even before they are ready to breed. In other species, like golden orb-weaver spiders (*Trichonephila clavipes*), males may spend a significant portion of their lives guarding females before and after they copulate with them.[20] In this species, mate guarding is also strongly influenced by the number of males in the area. In some cases, guarding by males prevents females from copulating with males they prefer, while in others, mate guarding may be welcomed by females because it prevents unwanted harassment from other males.

Mate guarding is an extensively studied topic, and there is no problem finding studies that report on guarding in different species. While humans make poor study subjects for various mating behaviours, there have been numerous investigations of mate guarding by both males and females in our species. Preventing unwanted attention from potential sexual competitors is often evident in social situations. One of the latest studies suggests that mate guarding by males increases when their mates are wearing high heels.[21] These shoes make a woman's legs look longer, deemed attractive by both males and females.

8.4 Sexual conflict during fertilization

Conflict can occur whenever the ability to select a mate is taken away from the female by the male's actions. If a male breeds with a female, there is no guarantee that his sperm will be of high quality or even fertilize the egg. Males usually have to compete against other males for the opportunity to copulate, but females often have the final say in determining which male is successful. This arrangement sets up a conflict between the sexes, and a highly active area of research concerns the battle between males and females to control fertilization. This conflict occurs in external fertilizers, where eggs may be subject to the sperm of many males, and internal fertilizers, where the female mates with more than one male (forcibly or consensually) during a given reproductive period.

There is always a period between the release of gametes and fertilization. This time is minimal in some species—it is practically nonexistent in many amphibians and fish, where male sperm is deposited directly onto females' eggs. In some cases, however, this period may be extensive, and sperm must last hours, days, or even longer in an aquatic medium or female reproductive tract before fertilization occurs. Although we often think of this as a relatively benign time (if we think of it at all), it can serve as a period during which either males or females may influence the success of gametes.

Much of the work regarding the control over fertilization has been conducted with sea urchins, a group for which syngamy is easy to replicate in the lab. There are almost a thousand species of sea urchins, but all breed sexually, and the males and females release an abundance of gametes during broadcast spawning. If all goes right, these meet up at the sea surface, where fertilization occurs. However, the female has to ensure that only one sperm gets through to fertilize the egg; if more than one sperm penetrates the egg membrane, a condition known as 'polyspermy', the egg and all of the sperm will be wasted. This outcome is not beneficial for anybody's reproductive success, particularly the females who have already expended more on the egg than males have on sperm. So, a male releases many sperm to better his chances of fertilizing the eggs, especially in competition with other males. On the other hand, the female produces an egg that makes it difficult for the sperm to get through. After all, the most vigorous sperm likely comes from a high-quality male, and that's the sperm she wants.

Gamete recognition proteins (GRPs) are known to be instrumental in the success of fertilization and have been well-studied in marine invertebrates.[22] These proteins, found on both the sperm and egg, permit the gametes to identify each other and for fertilization to occur. The compatibility of male and female proteins may vary depending on the chance of them coming together. If the fertilization rate is low, the compatibility of these proteins should be high to facilitate recognition. Conflict between the sexes is more likely to occur when there is an abundance of sperm and rates of polyspermy are high. Male sperm will always be pressed to quickly recognize female eggs—mainly if there is sperm from other males prowling around. Research on red sea urchins (*Strongylocentrotus franciscanus*), for example, suggests that GRPs become more critical as many males compete for fertilizations, as only those with appropriate matching GRPs will fertilize the egg.[23]

8.5 Postfertilization sexual conflict

In many species, care by one or both parents after the initial contribution to the gametes can improve the offspring's chances of survival (Figure 8.6). If the assistance of both parents is essential to the survival of the young, a relatively rare occurrence in the animal world, the choice is pretty simple: both parents must stick around and look after

Figure 8.6 *Elephants have one of the most prolonged periods of parental care of any animal. First, mothers carry the fetus for nearly two years before giving birth. After being born, calves will nurse for about six months, consuming about 10 L of milk their mother produces daily. Offspring will then supplement their diet with solid food but continue to obtain nutrition from their mother's milk. Young animals will continue to learn from their mother and other social group members for years.*

their offspring. Even so, there may still be conflict over the total amount of care each provides—the more responsibility with which you can burden your mate, the more energy you will have left to do other things, including finding another partner. This strategy may not seem very romantic, but we are talking about organisms we don't usually associate with candlelit dinners. Of course, it may apply to human parents as well.

The importance of the relative contribution of each of the parents in raising the offspring was a big question when I was a graduate student. I, and many of the researchers around me, worked on various species of birds; birds are one of the groups in which we often find substantial investment by both the male and female parent. The experimental approach to addressing this question was to remove one of the parents. I hate to admit that some of the investigators simply shot the male (a practice that would thankfully be frowned upon today), while the more scrupulous confined males to aviaries for the duration of the reproductive period. (Fortunately, I did not need to remove males as I was working on brown-headed cowbirds, a brood parasite in which neither parent cared for their offspring after egg-laying.) Other investigators have reduced the amount of care one parent provided (usually the male) by rendering care energetically more expensive or making the alternatives (e.g. seeking out other mates) easier.

Remember, the parent's contribution to care means that time and energy are being taken away from something else, something that will increase its future reproductive performance. In other words, each parent has to maximize its contribution to its current offspring (current fitness) while still investing in other activities that will increase its survival and ability to breed in the future (future fitness). Think of it as a trade-off between current and future reproductive bouts. What happens if one parent decreases the amount of care it provides? The other is faced with a problem and has four choices. It can (i) desert the current offspring altogether, (ii) invest slightly less (an amount similar to that of its partner), (iii) provide the same amount of care had its partner continued to invest the same amount, or (iv) invest slightly more to compensate for the decreased care of its mate. Unfortunately, there is no clear-cut prediction of what the individual should do. Several models suggest which of these would be the best strategy for maximizing reproductive output; most conclude that any change in parental care would depend on many factors, including the costs and benefits of decreased care and numerous ecological variables expected to change with each reproductive bout. In an examination of 53 studies of birds involving 33 species, one parent was either removed or manipulated so that the amount of parental care it provided was reduced.[24] As a result, the other parent indeed increased its care, but the total care was still less than that at nests with two parents. In other words, they either could not compensate for the reduced care by their mates entirely, or it wasn't in their best interest to do so.

In many species, only one parent provides care for the offspring. Clearly the sex that doesn't provide the care is better off. They gain by the increased survival of their present offspring (thanks to their partner), plus they are free to engage in other activities, which may include seeking other mating opportunities. Looking at it another way, the sex providing the care pays the costs and receives the benefits of the care, while the sex that leaves also gets the benefits but suffers none of the costs. This pattern of care sets up the perfect scenario for sexual conflict—if offspring require care by only one sex, which one will it be, the male or female?

Eurasian penduline tits (Figured 8.7) are most closely related to the North American verdins in the Southwestern United States. Never heard of verdins? Think of them as similar to the more common chickadees to which they are also related. They are named after their somewhat pendulous nests that typically hang over water. In this species, only one parent stays and looks after the nestlings, either the female (50%–70% of the time) or the male (5%–20%). The other parent generally deserts the nest before incubation of the eggs begins. In 20%–40% of all nests, both parents abandon the nest, resulting in the demise of all eggs. Not surprisingly, if the female deserts and

Figure 8.7 *A Eurasian penduline tit parent feeds its offspring. Only one parent is required to care for the offspring, and this might be the male or the female. Because of the potential benefits of nest desertion, both parents often abandon the nest, leading to clutch failure.*
Credit: Imagebroker.com/Shutterstock Photo ID 1,752,142,379.

the male stays, she benefits, and her caring partner suffers; if he deserts, the reverse is true. Of course, if both desert, they lose any of their present offspring, but at least both are free to pursue other mating prospects. The benefits of deserting, for both males and females, probably explain why biparental nest desertion is so high in this species.[25]

If one sex deserts the nest during the rearing of offspring, the remaining sex doesn't always suffer. We'd expect selection for behaviours in the remaining sex, which would reduce any possible costs arising from mate desertion. This situation may be the case in burying beetles. Although raising the young larvae is likely expensive, the female often benefits when males desert the nest. This is probably because, in the male's absence, the female can feed on the carcass generally reserved as a food source for the larva.[26]

When thinking about parental care, it is always helpful to determine which sex has the most to gain and which pays the highest cost of both staying and going. Of course, this will depend mainly on what your mate does. Thus, it becomes something of a guessing game between males and females. Unfortunately, those who repeatedly guess wrong won't leave many descendants.

8.6 Conflict outside of breeding

I've focussed on conflicts that arise between the sexes during breeding activities. However, males and females often live separate and dissimilar lives outside the breeding season. They may feed on different items, live in different areas, and exploit different niches. Such distinctions may mean that the evolutionary optima may differ for various anatomical and behavioural characteristics. For example, male birds may feed on larger seeds than females; thus, larger bills may be an advantage. But the same genes code for bill size in both males and females, so we have a tug-of-war—males are better off with larger bills, while females are better off with smaller bills. One of the first examples of this chapter involved the wing lengths of pied flycatchers. It is important to remember that while many differences between males and females arise because of sex-specific breeding strategies, these may also impact nonreproductive aspects of their lives.

8.7 Summary

Males and females often have different evolutionary interests, resulting in conflict at various points in the reproductive cycle. Genetically, this clash can be defined as intralocus or interlocus sexual conflict. 'Intralocus competition' refers to situations having different fitness optima for males and females for a characteristic determined at the same locus or loci. For example, the optimal size of males and females may differ, and we'd expect the genes for size to be the same in both sexes. Interlocus selection typically involves an interaction between the sexes, with the best outcome of this interaction being different for males and females. I subsequently look at examples of this conflict occurring before, during, and after fertilization. Before fertilization, it is possible to see the interests of the two sexes differ over nuptial gifts, copulatory plugs, forced copulation, the genitalia of males, and mate guarding. Conflict may subsequently involve the recognition proteins that allow gametes to identify each other during fertilization. If there needs to be further care for the offspring after fertilization, conflict can arise over the amount of care provided by each sex or, if only one parent is required, which cares. These are just a small sample of the potential conflicts between the sexes during a reproductive bout, and many others exist.

References

1. Schneider JM, Lubin Y. Infanticide by males in a spider with suicidal maternal care, *Stegodyphus lineatus* (Eresidae). Anim Behav [Internet]. 1997 Aug [cited 2022 Aug 20];54(2):305–312. Available from: https://doi.org/10.1006/anbe.1996.0454
2. Tarka M, Akesson M, Hasselquist D, Hansson B. Intralocus sexual conflict over wing length in a wild migratory bird. Am Nat [Internet]. 2014 Jan [cited 2022 Aug 20];183(1):62–73. Available from: https://doi.org/10.1086/674072
3. Li XW, Fail J, Shelton AM. Female multiple matings and male harassment and their effects on fitness of arrhenotokous *Thrips tabaci* (Thysanoptera: Thripidae). Behav Ecol Sociobiol [Internet]. 2015 Oct [cited 2023 Mar 16];69(10):1585–1595. Available from: http://www.jstor.org/stable/43599667
4. Kubli E. Sex-peptides: seminal peptides of the *Drosophila* male. Cell Mol Life Sci [Internet]. 2003 Aug [cited 2022 Aug 22];60:1689–1704. Available from: https://doi.org/10.1007/s00018-003-3052
5. Albo MJ, Winther G, Tuni C, Toft S, Bilde T. Worthless donations: Male deception and female counter play in a nuptial gift-giving spider. BMC Evol Biol [Internet]. 2011 [cited 2023 Mar 15];11:329–336. Available from: https://doi.org/10.1186/1471-2148-11-329
6. Bussière LF, Hunt J, Jennions MD, Brooks R. Sexual conflict and cryptic female choice in the black field cricket, *Teleogryllus commodus*. Evolution [Internet]. 2006 Apr [cited 2023 Mar 16];60(4):792–800. Available from: http://www.jstor.org/stable/4095294
7. Mangels R. Tsung K, Kwan K, Dean MD. Copulatory plugs inhibit the reproductive success of rival males. J Evol Bio [Internet]. 2016 Nov [cited 2023 Aug 30]; 29(11):2289–2296. Available from: https://doi.org/10.1111/jeb.12956
8. Tarpy DR, Delaney DA, Seeley TD. Mating frequencies of honey bee queens (*Apis mellifera* L.) in a population of feral colonies in the Northeastern United States. PloS One [Internet]. 2015 Mar [cited 2023 Mar 16];10(3):e0118734. Available from: https://doi.org/10.1371/journal.pone.0118734
9. Baer B, Schmid-Hempel P. Experimental variation in polyandry affects parasite loads and fitness in a bumble-bee. Nature [Internet]. 1999 Jan [cited 2022 26 Aug];397:151–154. Available from: https://doi.org/10.1038/16451
10. Dukas R, Jongsma K. Costs to females and benefits to males from forced copulations in fruit flies. Anim Behav [Internet]. 2012 Nov [cited 2023 Mar 16];84(5):1177–1182. Available from: https://doi.org/10.1016/j.anbehav.2012.08.021
11. Brennan PL, Gereg I, Goodman M, Feng D, Prum RO. Evidence of phenotypic plasticity of penis morphology and delayed reproductive maturation in response to male competition in waterfowl. Auk [Internet]. 2017 Sep [cited 2022 Aug 27];134(4):882–893. Available from: https://doi.org/10.1642/AUK-17-114.1
12. Adler M. Sexual conflict in waterfowl: Why do females resist extrapair copulations? Behav Ecol [Internet]. 2010 Jan/Feb [cited 2022 Mar 16];21(1):182–192. Available from: https://doi.org/10.1093/beheco/arp160
13. Han CS, Jablonski PG. Female genitalia concealment promotes intimate male courtship in a water strider. PLoS One [Internet]. 2009 Jun [cited 2023 Mar 19];4:e5793. Available from: https://doi.org/10.1371/journal.pone.0005793
14. Han CS, Jablonski PG. Male water striders attract predators to intimidate females into copulation. Nat Commun [Internet]. 2010 Aug [cited 2023 Mar 19];1(5):1–6. Available from: https://doi.org/10.1038/ncomms1051

15. Morrow EH, Arnqvist G. Costly traumatic insemination and a female counter-adaptation in bed bugs. Proc Biol Sci [Internet]. 2003 Nov [cited 2023 Mar 19];270(1531):2377–2381. Available from: https://doi.org/10.1098/rspb.2003.2514
16. Hotzy C, Arnqvist G. Sperm competition favors harmful males in seed beetles. Curr Biol [Internet]. 2009 Feb [cited 2022 Aug 26];19:404–407. Available from: https://doi.org/10.1016/j.cub.2009.01.045
17. Zhang Z, Head M. Does developmental environment affect sexual conflict? An experimental test in the seed beetle. Behav Ecol [Internet]. 2022 Jan/Feb [cited 2022 Aug 26];33(1):147–155. Available from: https://doi.org/10.1093/beheco/arab119
18. Friesen CR, Uhrig EJ, Squire MK, Mason RT, Brennan PL. Sexual conflict over mating in red-sided garter snakes (*Thamnophis sirtalis*) as indicated by experimental manipulation of genitalia. Proc Biol Sci [Internet]. 2013 Nov [cited 2022 Aug 26];281(1774):20132694. Available from: https://doi.org/10.1098/rspb.2013.2694
19. Rueda-Solano LA, Vargas-Salinas F, Pérez-González JL, Sánchez-Ferreira A, Ramírez-Guerra A, Navas CA, et al. Mate-guarding behaviour in anurans: intrasexual selection and the evolution of prolonged amplexus in the harlequin toad *Atelopus laetissimus* Anim Behav [Internet]. 2022 Mar [cited 2023 Apr 8];18:127–142. Available from: https://doi.org/10.1016/j.anbehav.2021.12.003
20. Del Matto LA, Macedo-Rego RC, Santos ES. Mate-guarding duration is mainly influenced by the risk of sperm competition and not by female quality in a golden orb-weaver spider. PeerJ [Internet]. 2021 Oct [cited 2023 Apr 8];9:e12,310. Available from: https://doi.org/10.7717/peerj.12310
21. Prokop P. High heels enhance perceived sexual attractiveness, leg length and women's mate-guarding. Curr Psychol [Internet]. 2022 May [cited 2023 Apr 13];41(5):3282–3292. Available from: https://doi.org/10.1007/s12144-020-00832-y
22. Levitan DR. Do sperm really compete and do eggs ever have a choice? Adult distribution and gamete mixing influence sexual selection, sexual conflict, and the evolution of gamete recognition proteins in the sea. Amer Nat [Internet]. 2018 Jan [cited 2023 Mar 19];191(1):88–105. Available from: http://purl.flvc.org/fsu/fd/FSU_libsubv1_wos_000418197900009
23. Harrison F, Barta Z, Cuthill I, Székely T. How is sexual conflict over parental care resolved? A meta-analysis. J Evol Biol [Internet]. 2009 Sep [cited 2022 Aug 26];22(9):1800–1812. Available from https://doi.org/10.1111/j.1420-9101.2009.01792.x
24. Pogány Á, Kosztolányi A, Miklósi Á, Komdeur J, Székely T. Biparentally deserted offspring are viable in a species with intense sexual conflict over care. Behav Proc [Internet]. 2015 Jul [cited 2023 Mar 19];116:528–32. Available from: https://doi.org/10.1016/j.beproc.2015.04.014
25. Boncoraglio G, Kilner RM. Female burying beetles benefit from male desertion: Sexual conflict and counter-adaptation over parental investment. PLoS One [Internet]. 2012 Feb [cited 2022 Aug 26];7(2):e31713. Available from: https://doi.org/10.1371/journal.pone.0031713
26. Vacquier VD, Swanson WJ. Selection in the rapid evolution of gamete recognition proteins in marine invertebrates. Cold Spring Harb Perspect Biol [Internet]. 2011 Nov [cited 2023 Mar 16];3(11): a002931. Available from: https://doi.org/10.1101/cshperspect.a002931

9

Co-operation Between the Sexes

9.1 When do we expect co-operation?

Kirk's dik-diks (*Madoqua kirkii*) are dwarf antelopes commonly seen on African safaris (Figure 9.1). They are almost always observed in pairs. If you look closely, these pairs consist of males and females; the females are slightly larger, and the males have small horns. Oddly, for a mammal, they're monogamous and remain together throughout their lives. They can breed at 6–12 months, produce a single offspring every 6 months, and live up to 10 years. This means that they will often have nearly 20 reproductive periods. There is no evidence of extrapair copulations, and the two individuals remain together throughout the year, spending, on average, almost two-thirds of their time together.[1] Their reproductive behaviour means that males and females of a pair have similar reproductive interests. Both will maximize their fitness by producing many high-quality offspring. It's difficult to imagine sexual conflict as underlying their entire relationship. Indeed, there must be some co-operation if males and females are to raise offspring and maximize their reproductive goals.

In terms of natural selection, it pays to be selfish. But do males and females never co-operate to improve their reproductive performance? Co-operation between the members of a breeding pair is related to the overlap in their genetic interests. In other words, males and females will co-operate if it is in their best interest to do so. If the dik-dik parents produce offspring together for most of their reproductive lives, the genetic interests are similar, conflict would be significantly reduced, and co-operation would increase. If, alternatively, a male could increase his reproductive output by breeding with another female, and his mate does better having him around, conflict between the sexes will increase. The degree of conflict or co-operation depends mainly on the intensity of sexual selection; in general, conflict between males and females will intensify with increasing competition between members of one sex for the other.

However, co-operation, as a reproductive strategy, occurs not only between parents but also between same-sex individuals. Males may co-operate with other males, or females will form alliances with other females to improve their reproduction success. These are also important reproductive strategies, and they will be reviewed here. It is helpful to understand the conditions under which individuals of each sex work together

The Evolution of Sex. Kevin Lee Teather, Oxford University Press. © Kevin Lee Teather (2024).
DOI: 10.1093/9780191994418.003.0009

Figure 9.1 *Dik-dik pairs are extremely loyal. This male and female from Masai Mara in Kenya will probably mate for life, going through about 20 reproductive bouts. Having similar reproductive interests encourages a great deal of co-operation between the sexes.*
Credit: PhotocechCZ/Shutterstock Photo ID: 468,698,534.

because it provides a better idea of when co-operation is beneficial. Recall that the reproductive success of males is usually limited by the number of eggs they can fertilize; these are often produced by different females. Conversely, females' reproductive success is most often determined by the number and quality of offspring they can generate. Naturally, there are exceptions to both of these. Therefore, we would expect males to co-operate if it better allows them to obtain and defend resources needed by females or the females themselves. Females, on the other hand, would be more likely to co-operate if it decreases the costs of manufacturing large gametes and raising offspring.

Direct fitness payoffs can explain much of the co-operation we see, especially in mated pairs with common genetic interests. With direct fitness gains, both members of the co-operating pair gain because of the combined actions of the two. In same-sex individuals, such direct benefits are often immediate for one and delayed for another, as will be clarified below. However, the co-operation between individuals can also evolve because of, or be augmented by, **indirect fitness** gains. William Hamilton, who we've come across before (remember, he was the one who championed the idea that sexual reproduction is a strategy to combat pathogens), may be best known for his work on altruism. Hamilton pointed out that there are two ways for an organism to get genes into the next generation.[2] It could do it directly by having its own offspring or indirectly by promoting its genes in close relatives. J. B. S. Haldane reportedly stated that he would jump into a river to save two brothers or eight cousins. Hamilton clarified the reason: You share half of your genes, by common descent, with your brother and one-eighth with your cousin. Hamilton's rule states that an altruistic behaviour could evolve if the benefit derived by the recipient (in terms of fitness) multiplied by the degree of relationship between the individuals is greater than the cost of that behaviour accrued by the performer:

$$rB > C$$

Hamilton's rule can explain the co-operation observed in closely related individuals, including those in ant colonies (remember that ants are haplodiploid—this means that sisters are actually related to each other by three-quarters as opposed to one half). The sum of direct and indirect fitness is referred to as **inclusive fitness**.

Both direct and indirect fitness can explain co-operation within the sexes (intrasexual co-operation) or between sexes (intersexual co-operation), usually the members of a breeding pair. As we would expect males and females to be under different selective pressures, the factors leading to the evolution of co-operation would differ for each. Let's begin by looking at examples where individuals of the same sex co-operate. Remember, interactions between individuals are expected to have elements of both conflict and co-operation, making them relatively complex.

9.2 Intrasexual co-operation

Although we generally expect individuals within one sex to compete for opportunities to mate with the other, this is not always the case. We sometimes find males co-operating to increase their fitness. Similarly, more than one female can collaborate in a way that will maximize their reproductive performance.

9.2.1 Between males

While sitting on the top of a passenger boat travelling down a river in Borneo one night, I was fortunate enough to witness an amazing spectacle. Entire trees were lighting up, pulsating every second or so, in a display that could be seen for miles (Figure 9.2). Having studied animal behaviour at university, I understood that these were all male fireflies, synchronizing their flashing pattern to make females more likely to see and recognize them. Leks usually consist of groups of males clustered in a small area and are discussed more fully later. However, in most lek mating systems, males display together as a primary strategy for getting a mate. Sure, there may be other strategies at play. Quiet males may hang around the periphery of leks with the hope of intercepting and copulating with females. To my knowledge, such alternative mating tactics have never been considered in these mating swarms of fireflies, and they might be difficult to examine. But generally,

Figure 9.2 *In some firefly species, males time their flashes to coincide with others around them. Presumably, this brighter signal attracts more females to mate, and participating males potentially benefit. The trees are so bright that fishermen along the coasts have used them as a guide to specific rivers.*

Credit: Footage Lab/Shutterstock Photo ID 1,063,564,289.

the synchronization of flashing by fireflies can be viewed as a form of co-operation; males expend a certain amount of energy to increase the probability of mating for themselves (intentionally) and those around them (unintentionally); those that flash synchronously are more likely to breed.

In general, the displays by groups of males (sometimes genetically unrelated, but often we don't know) may be instrumental in attracting and copulating with females. In other species, co-operating males may gain experience to make them more successful in the future. In lance-tailed manakins (*Chiroxiphia linearis*), two males co-operate to perform a courtship in the hope of attracting a female. They sing and undergo a relatively complex courtship dance, often remaining together for more than one year. However, a hierarchy exists among the males, presumably based on age, and only the alpha male mates. The other, referred to as the beta male, is unrelated to his partner, so he gains no indirect benefits through kin selection. Instead, he gains the skills necessary to become an alpha male in the future. Indeed, males who serve as apprentices are more likely to eventually become the alpha members of a pair than males who don't.[3] Thus, the fitness benefits of co-operating with another male to perform elaborate courtship rituals are delayed.

Alternatively, kin selection may promote co-operation by providing indirect fitness benefits. For instance, about 30% of male dunnocks (*Prunella modularis*) who breed with a single female (the variable mating system of the dunnock is discussed more fully in Section 9.3.1) are more likely to be related so that they receive indirect and direct benefits.[4] Wild turkey (*Meleagris gallopavo*) males establish small coalitions of two or three males within larger groups, defending females from other such groups. However, only one male copulates with any females attracted. The help provided by the other male(s) increases the reproductive success of the breeding male, above that which could have been accomplished with no helpers.[5] So, how do helpers benefit? This breeding system provides an excellent example of kin selection, as the helpers are closely related to the breeding male genetically. Based on the similarity of microsatellite DNA, the average degree of relatedness (r in Hamilton's equation above) was 0.42. Full siblings have an r of 0.5. Thus, co-operation among male turkeys is a beneficial strategy that has likely evolved through kin selection.

Co-operating individuals often gain from both direct and indirect fitness benefits. We've looked at lion prides previously, but let's go into their social organization in a little more detail as they may serve to illustrate this point. Once males are old enough, they leave the natal pride, seeking another group of females to take over from the resident adult males. Initially, these groups of roaming males can be anywhere from 1–7 individuals, but single or smaller groups may join up with one or more unrelated males before attempting a pride takeover. Therefore, coalitions of males are typically related (brothers or cousins born in the same pride) but can be genetically unrelated, as is the case when small groups of males join to form a coalition. Being in a larger group of males has its advantages. They are more likely to take over a pride of females, better defend the territory and the females, remain residents in the pride longer, and increase their reproductive success.[6,7]

9.2.2 Between females

Female reproductive success is usually not limited by mate availability. Instead, the success of females is generally limited by how many high-quality eggs they can produce or offspring they can rear. This difference means that while males form coalitions that enable them to increase mating opportunities, females will co-operate if it provides them with better access to resources required to produce gametes or, as in the case of lions, better protection for their offspring.[7] Females will also collaborate outside the mating season to increase social bonds and presumably the group's cohesiveness. Ring-tailed coatis (*Nasua nasua*) are a raccoon-like mammal found in the forests of South America. The females live in social groups where individuals engage in several co-operative activities, including grooming, nursing, predator defence, and support in aggressive interactions with other band members (Figure 9.3). While coati females within a group are suspected of being related, co-operation between females might result from direct fitness benefits (in the form of reciprocal altruism) in addition to kin selection. As expected, females who assist others are more likely to be helped in the future.[8]

Figure 9.3 *Ring-tailed coatis are very social, and the females co-operate in many activities. The females are likely related, meaning they get indirect fitness benefits by helping close relatives. However, females who help now are more likely to be helped in the future, suggesting they may also receive delayed direct fitness benefits.*

Credit: Rudiernst/Adobe Photo ID 578,135,612.

Raising offspring communally with other females may be an excellent way to decrease the costs involved in parental care. Indeed, communal care has been documented in 15% of mammals and 2.5% of birds. In house mice (*Mus musculus domesticus*), females adopt one of two breeding strategies: raise their brood alone or combine it with that of another female. In the latter case, both females protect and provide milk indiscriminately to all offspring, whether they belong to them or not. Unlike solitary nesters, the amount of milk they produced and fed was not based on the number of offspring to which they gave birth, but rather the combined number of offspring in both litters.[9] Thus, co-operatively nursing females do not derive equal benefits regarding the amount of milk fed per pup unless litter sizes are identical for both. This scenario presents an opportunity for exploitation. However, if the co-operative nursers are related, the cost of overfeeding decreases. As it turns out, females are more likely to nest communally if partners come from a pool of genetically related individuals.[10] On the other hand, it's possible that communal nesting in this species may simply come down to being the best option at the time, given environmental constraints,[11] such as food availability.

I think everyone has heard of bonobos and their active sex lives. Bonobos are one of two kinds of chimpanzees but have social lives quite dissimilar from their close relatives.

Sex in this species is thought to alleviate tension, and aggressive interactions are much less frequent in bonobos than in the troops of their chimp cousins. 'Make love, not war' seems to be their mantra. Sexual activity takes place between females and other females, males and males, and even between youngsters, sometimes with each other and sometimes with adults. And, oh yeah, adult males and females engage in sex too. Besides reducing tension in the group, sex is thought to promote co-operation, especially among females. Indeed, sexual activity between females, involving genito-genital rubbing, was much more frequent than sex between males and females.[12] Moreover, it increased the oxytocin levels of females, a hormone that promotes co-operation, a result not observed in females engaging in sex with males.

9.3 Intersexual co-operation

Although males and females have their own breeding agendas, they often must work cooperatively to carry out certain tasks. Usually, the two individuals are not closely related and each gain more fitness benefits by working together than by acting alone.

9.3.1 Biparental care: Conflict and co-operation

Biparental care occurs when two parents combine their efforts to raise offspring successfully. This type of care is seen in most birds as well as some mammals, amphibians, fish, and arthropods. While the young of most teleost fish, for example, are cared for by one parent, the cichlid family is an exception, with 42% of the 1700 described species exhibiting biparental care.[13] In all biparental species, a certain amount of co-operation between males and females is inevitable since they have overlapping genetic interests best achieved by collaboration. However, we have previously seen that conflict can result in the amount of care each provides. This amount will always depend on the costs and benefits to each parent of delivering care to the current offspring (Figure 9.4). The benefits are straightforward: any care given to the young increases the probability that they will survive and raises the fitness of both parents to the extent that the offspring belong to them. The costs, on the other hand, are numerous. They involve not only the time, energy, and risk involved in caring for the offspring, but also in lost breeding opportunities. Since males are more likely to increase their reproductive output by mating with other females, we would expect that the loss of such opportunities would be more detrimental. Therefore, biparental care can be considered a balance between co-operation and conflict, with the amount of care provided by males and females dependent on the advantages and disadvantages that each face.

Of course, two individuals collaborating to raise offspring isn't the same as two individuals doing the same things, and the two sexes' activities are often task-specific. For example, males may contribute more to defence; females more to incubating and brooding; and both parents equally to feeding offspring. Thus, it is difficult to compare the costs to each parent in rearing their progeny. However, the potential fitness benefits to

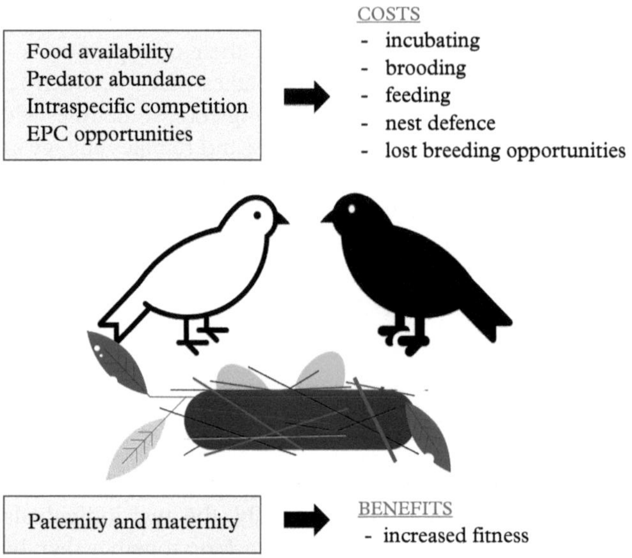

Figure 9.4 *There are many costs and benefits of biparental care. Not only is there time and energy invested in activities such as caring directly for the offspring and nest defence, but males and females have less time to be involved in extrapair copulations. These may differ for each parent. The fitness benefits also may differ as they depend on the certainty of parentage. Females are nearly always confident that they are the mother; males of internal breeders are rarely certain.*

each are quantifiable as long as we know the genetic relationship between the offspring and their parents. We can usually be reasonably confident that all the offspring in the brood belong to the female (although this certainty may be reduced in cases of brood parasitism or if fertilization is external). In his own nest, male fitness benefits may vary with the level of extrapair fertilizations in the population. However, fitness will increase if he fertilizes females who are not his mate. These sex-specific costs and benefits can change depending on environmental conditions.

Let's examine two factors influencing these costs and benefits to demonstrate how co-operation between males and females in the biparental care of offspring might be impacted. The first is extrapair copulation opportunities. The opportunity to fertilize a female who is not your mate will depend on many variables, especially the receptivity of other females and competition with other males. It will also impact paternity certainty in a male's nest. To make it even more complicated, the success of EPCs is likely dependent on the male's quality—both in his attractiveness to other females and his ability to limit breeding opportunities available to his partner. The second factor that may influence the costs and benefits of co-operation is food abundance; as we will see, the effect of food abundance may also depend on how food is provided to offspring. Nothing is straightforward!

Hair-crested drongos (*Dicururus hottentottus*) are an Asian bird species best known for their nest-dismantling behaviour. Immediately after nesting, the adults (probably both males and females) pull apart their nests. The material isn't reused, and the behaviour is thought to reduce predation risk for the fledglings.[14] In this monogamous species, both males and females feed the nestlings. Given their limited energy supply, high-quality males can put their resources into feeding offspring or seeking extrapair fertilizations, whichever maximizes their fitness. Since there is such a significant fitness payoff to fertilizing your female neighbour while not suffering any of the costs of rearing offspring, the choice is obvious. Thus, male hair-crested drongos who engage in extrapair copulations do not contribute as muh to raising their own offspring,[15] thereby reducing co-operation with their mate. Not surprisingly, though, the frequency of extrapair copulations depends on the availability and receptivity of other females.

Perhaps the best evidence of paternal parental care and fitness benefits comes from the classic studies by Nick Davies on the dunnock, a small garden bird found throughout Europe.[16,17,18] Dunnocks have a variable mating system where some birds breed in pairs (monogamy), some females have multiple male partners (polyandry), some males have numerous female mates (polygyny), and some mating groups consist of multiple females and males (polygynandry; Figure 9.5). As you would imagine, both conflict and co-operation can be at play in a population of dunnocks,[4] but let's look specifically at paternity. A polyandrous female solicits copulations from two males who both fertilize a percentage of her eggs. As a result, a male will adjust the amount he feeds nestlings to correspond with the number of offspring in the nest that belong to him. How does he know what percentage is his? This percentage closely aligns with his copulation rate relative to the other male. Therefore, he invests energy in parenthood to match his success at copulating with the female.

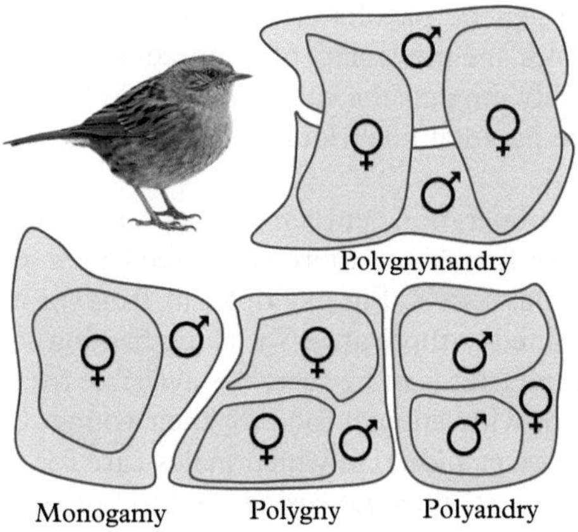

Figure 9.5 *Dunnocks have variable mating systems. They can mate monogamously, polyandrously, polygynously, or even polygynandrously. The many ways males and females can breed depend on the options available and highlight the interplay between conflict and co-operation when biparental care is necessary.*

Dunnock photo credit: Shutterstock Photo ID 738,361,765.

Evaluating paternity in dunnocks is relatively simple; females don't try to hide their copulation activity, and males can estimate the number of offspring to which they are genetically related. In other species, males may adjust their paternity to the threat of being cuckolded. House wrens (*Troglodytes aedon*) exhibit this strategy; males reduce their parental care in response to a perceived chance of another male fertilizing his mate.[19] However, determining the possibility of EPCs is sometimes more difficult as females are often secretive about their rendezvous with other males. For example, female thorn-tailed rayaditos (*Aphrastura spinicauda*), a monogamous bird with biparental care found in the southern regions of South America, are much more discrete. Males presumably can't assess the number of offspring they sire, and there is no evidence that paternity levels affect the amount of care they provide.[20] Additionally, neither the threat of EPCs nor the actual proportion of offspring that belong to them influences the provisioning rate in grass wrens (*Cistothorus platensis*).[21]

Thus, the amount of care a male might devote to his offspring depends on the time spent obtaining extrapair fertilizations. But what influences the availability of EPCs? In birds, at least, females should control the frequency of EPCs as they must usually co-operate with the males for copulation to occur. Females can engage in extrapair fertilizations for various reasons, including (i) the possibility that her mate didn't fertilize all her eggs, (ii) increasing the genetic variability of her offspring, (iii) copulating with a higher-quality male than her mate, or (iv) mating with a more distantly related male to reduce the potential impact of inbreeding depression. For example, Japanese quail (*Coturnix japonica*) females may base their availability for EPCs on the success of their own mate in fertilizing their eggs previously.[22] However, as she can be relatively sure (at least in species that fertilize internally) that all the offspring are hers, regardless of who fathers the brood, she will continue to maximize her care to the extent that future fitness benefits aren't compromised.

While the overlap in genetic payoffs affects the level of conflict and co-operation between males and females, the availability of resources can also impact parental care. In particular, food availability impacts the cost-benefit ratio for each sex. However, abundant food can influence parental behaviour in different ways, particularly the options open to males.

In birds, the level of co-operation between parents is expected to decrease with abundant food as it will allow females to more easily feed their young, freeing up males for other activities, such as EPCs. For example, in frugivorous species, the range of female-only care is restricted to those areas with long fruiting seasons.[23] Biparental care is necessary in areas where fruit is more scarce or available for a shorter period, as both parents are required to provide enough food for their young. Compare this with burying beetles (*Nicrophorus vespilloides*), in which males care for their brood longer when there is more food at the nest, thus increasing the level of co-operation.[24] The difference between these two lies in how the food is made available to progeny. For birds, parents must forage and return to the nest with food for nestlings. This activity can be extremely costly, and any reduction in feeding trips will free up time and energy. Burying beetles, on the other hand, bury an entire animal corpse and lay eggs around it. Both parents regurgitate food from the carcass to the larvae after they hatch. Parents also use the carcass as a food source for themselves. Thus, instead of expending a great deal of energy obtaining food for their young, larger carcasses reduce the costs of feeding young, allowing males to gain weight over the reproductive cycle.

9.3.2 Are indirect fitness benefits significant?

Generally, we would expect co-operation between the male and female of a breeding pair to evolve because of direct benefits rather than kin selection. Inbreeding typically results in lower-quality and fewer offspring, thereby reducing parents' fitness and favouring mate selection that discriminates against relatives. However, there are also benefits to mating with closely related individuals. A higher percentage of an adult's genes may be passed on to offspring if parents are more closely related. As a result, kin selection can increase the amount of co-operation between individuals. Despite the potential costs, there are a few species in which mated pairs are close relatives. Is there any evidence that these are more likely to co-operate in raising offspring?

Nigerian reds (*Pelvicachromis taeniatus*; Figure 9.6) are cichlid fish that inhabit freshwater in and around Nigeria and Cameroon. Both males and females raise the brood together, with females caring for eggs and males defending the territory. Presumably, because of males' elevated level of parental investment, females compete for access to the best mates. *Pelvicachhromis taeniatus* is also one of the few species where breeding partners have been reported to be closely related.[25] In mate choice experiments,

Figure 9.6 *A wild-caught male (below) and female Nigerian red in an aquarium. Pairs are often genetically related individuals, and thus, they get the benefit of having offspring that are more closely related to them. Furthermore, closely related pairs make better parents.*
Credit: Neale Monks/CC BY-SA 3.0.

both sexes preferred kin over nonkin as mates. Further, closely related individuals of a pair were better parents. Compared to outbred pairs (those who weren't related), kin-based pairs defended the territory more and spent significantly more time with free-swimming young, and adult males didn't attack females as often as outbred males did after the young swam free. Significantly, there was no inbreeding depression found for this species.

The importance of kin selection in promoting co-operation in a breeding pair depends on the strength of inbreeding depression and the intensity of sexual selection. When the cost of mating with a close relative is high, inbreeding depression may outweigh any advantage of increased co-operation. When inbreeding depression is low, as in Nigerian reds, the fitness gains through co-operation may evolve. Thus, **optimal outbreeding** may influence the degree of relatedness between reproductive partners.[26] In other words, it's best to mate with relatives (to obtain the indirect benefits associated with kin selection) but not overly close relatives (to avoid the costs of inbreeding). This trade-off is expected to vary by species.

Finally, the genetic relationship between the male and female of a breeding pair may be largely unknown, so it's difficult to assess the strength of kin selection in the evolution of intersexual co-operation. Great frigatebirds (*Fregata minor*) are long-lived seabirds that don't start breeding until they are about 10. While pairs don't mate for life, they have one of the most prolonged periods of parental care known in any bird and look after offspring for more than a year. Males and females both incubate the eggs and feed the young, suggesting substantial co-operation. Interestingly, in many cases, the breeding pair is more closely related than expected by chance, with the relationship being just less than that between first cousins.[27] Females actively select males that are closely related to themselves. Although this likely promotes co-operation between the pair, it may have other unknown advantages. Further work is needed to assess the relationship between mates, evaluating the degree of collaboration and the strength of inbreeding depression.

9.4 Summary

One or more of the following must be true in co-operating with another individual. First, co-operative behaviour is rewarded by **current direct fitness** benefits; thus, the cost of performing the behaviour is outweighed by the benefits of more than one individual doing it. For example, two individuals may be better than one at defending the resources needed for reproduction, and by doing so, they can reproduce at a higher rate than they would alone. Significantly, the costs don't rise as rapidly as the benefits. Sometimes, these direct benefits may be delayed such that present costs are outweighed by potential benefits in the future. In this way, a co-operator benefits from **future direct fitness**. Third, **indirect fitness** benefits exist where a greater proportion of your genes survive if you help a close relative. Such strategies can evolve through kin selection.

Intrasexual co-operation can serve a variety of functions but must result in an increase in fitness for the participants. Most often, male-male collaboration can lead to a better defence of resources needed by females or the protection of the females themselves. Females can reduce the costs of producing or rearing high-quality offspring by collaborating with other females. **Intersexual co-operation** is seen most commonly between members of a breeding pair. If biparental care is required to raise offspring successfully, and the genetic interests of males and females overlap substantially, parents must collaborate during the brood-rearing period. Several factors, such as the availability of EPCs and food, can influence the costs and benefits of co-operation for each sex. It is important to remember, though, that such relationships involve both conflict and co-operation, and individuals will always try to maximize their own fitness.

References

1. Brotherton PN, Pemberton JM, Komers PE, Malarky G. Genetic and behavioural evidence of monogamy in a mammal, Kirk's dik-dik (*Madoqua kirkii*). Proc Biol Sci [Internet]. 1997 May [cited 2023 Mar 30];264(1382):675–681. Available from: http://www.jstor.org/stable/51087
2. Hamilton WD. The genetical evolution of social behaviour. II. J Theor Biol [Internet]. 1964 Jul [cited 2023 Mar 28];7(1):17–52. Available from: https://doi.org/10.1016/0022-5193(64)90039-6
3. DuVal EH. Adaptive advantages of cooperative courtship for subordinate male lance-tailed manakins. Am Natur [Internet]. 2007 Apr [cited 2023 Mar 30];169(4):423–432. Available from: https://doi.org/10.1086/512137
4. Santos ES, Santos LL, Lagisz M, Nakagawa S. Conflict and cooperation over sex: The consequences of social and genetic polyandry for reproductive success in dunnocks. J Anim Ecol [Internet]. 2015 Aug [cited 2023 Apr 3];84(6):1509–1519. Available from: https://doi.org/10.1111/1365-2656.12432
5. Krakauer A. Kin selection and cooperative courtship in wild turkeys. Nature [Internet]. 2005 Mar [cited 2023 Mar 31];434:69–72. Available from: https://doi.org/10.1038/nature03325
6. Packer C, Pusey AE. Intrasexual cooperation and the sex ratio in African lions. Am Natur [Internet]. 1987 Oct [cited 2023 Mar 24];130(4):636–642. Available from: https://doi.org/10.1086/284735
7. Grinnell J. Modes of cooperation during territorial defense by African lions. Hum Nat [Internet]. 2002 Mar [cited 2022 Mar 24];13(1):85–104. Available from: https://doi.org/10.1007/s12110-002-1015-4
8. Romero T, Aureli F. Reciprocity of support in coatis (*Nasua nasua*). J Comp Psychol [Internet]. 2008 Feb [cited 2023 Mar 27];122(1):19–25. Available from: https://doi:10.1037/0735-7036.122.1.19
9. Ferrari M, Lindholm AK, König B. The risk of exploitation during communal nursing in house mice, *Mus musculus domesticus*. Anim Behav [Internet]. 2015 Dec [cited 2023 Mar 25];110:133–143. Available from: https://doi.org/10.1016/j.anbehav.2015.09.018
10. Harrison N, Lindholm AK, Dobay A, Halloran O, Manser A, König B. Female nursing partner choice in a population of wild house mice (*Mus musculus domesticus*). Front Zool [Internet]. 2018 Mar [cited 2023 Mar 20];15:4. Available from: https://doi:10.1186/s12983-018-0251-3
11. Ferrari M, Lindholm AK, Ozgul A, Oli MK, König B. Cooperation by necessity: condition- and density-dependent reproductive tactics of female house mice. Commun Biol [Internet].

2022 Apr [cited 2023 Mar 25];5:348. Available from: https://doi.org/10.1038/s42003-022-03267-2

12. Moscovice LR, Surbeck M, Fruth B, Hohmann G, Jaeggi AV, Deschner T. The cooperative sex: Sexual interactions among female bonobos are linked to increases in oxytocin, proximity and coalitions. Horm Behav [Internet]. 2019 Nov [cited 2023 Mar 25];116:104581. Available from: https://doi.org/10.1016/j.yhbeh.2019.104581
13. Balshine S, Abate ME. Parental care in cichlid fishes. In: Abate ME, Noakes DL, editors. The behavior, ecology and evolution of cichlid fishes. Fish & Fisheries Series, vol 40. Dordrecht: Springer; 2021. p. 541–586. Available from: https://doi.org/10.1007/978-94-024-2080-7_15
14. Li J, Lin S, Wang Y, Zhang Z, Nest-dismantling behavior of the hair-crested drongo in Central China: An adaptive behavior for increasing fitness? Condor [Internet]. 2009 Feb [cited 2023 Mar 28];111(1):197–201 Available from: https://doi.org/10.1525/cond.2009.080051
15. Lv L, Zhang Z, Groenewoud F, Kingma SA, Li J, van der Velde M, et al. Extrapair mating opportunities mediate parenting and mating effort trade-offs in a songbird. Behav Ecol [Internet]. 2020 Mar/Apr [cited 2023 Mar 28];31(2):421–431 Available from: https://doi.org/10.1093/beheco/arz204
16. Davies NB. Cooperation and conflict among dunnocks, *Prunella modularis*, in a variable mating system. Anim Behav [Internet]. 1985 May [cited 2023 Apr 3];33(2):628–648. Available from: https://doi.org/10.1016/S0003-3472(85)80087-7
17. Davies NB, Houston AI. Reproductive success of dunnocks, *Prunella modularis*, in a variable mating system. II. Conflicts of interest among breeding adults. J Anim Ecol [Internet]. 1986 Feb [cited 2023 Apr 3];55(1):139–154. Available from: https://doi.org/10.2307/4698
18. Davies NB, Hatchwell BJ, Robson T, Burke T. Paternity and parental effort in dunnocks *Prunella modularis*: How good are male chick-feeding rules? Anim Behav [Internet]. 1992 May [cited 2023 Apr 3];43(5):729–745. Available from: https://doi.org/10.1016/S0003-3472(05)80197-6
19. DiSciullo RA, Thompson CF, Sakaluk SK. Perceived threat to paternity reduces likelihood of paternal provisioning in house wrens. Behav Ecol [Internet]. 2019 Sep/Oct [cited 2023 Mar 28];30(5):1336–1343. Available from: https://doi.org/10.1093/beheco/arz082
20. Poblete Y, Botero-Delgadillo E, Espíndola-Hernández P, Südel G, Vásquez RA. Female extrapair behavior is not associated with reduced paternal care in thorn-tailed rayadito. Ecol Evol [Internet]. 2021 Mar [cited 2023 Mar 28];11(7):3065–3071. Available from: https://doi.org/10.1002/ece3.7232
21. Arrieta RS, Campagna L, Mahler B, Liambías PE. Neither paternity loss nor perceived threat of cuckoldry affects male nestling provisioning in grass wrens. Behav Ecol Sociobiol [Internet]. 2022 Oct [cited 2023 Mar 28];76:147. Available from: https://doi.org/10.1007/s00265-022-03253-y
22. Vedder O. Experimental extrapair copulations provide proof of concept for fertility insurance in a socially monogamous bird. Proc R Soc B [Internet]. 2022 Jun [cited 2023 Mar 28];289(1976); 20220261. Available from: http://doi.org/10.1098/rspb.2022.0261
23. Barve S, La Sorte FA. Fruiting season length restricts global distribution of female-only parental care in frugivorous passerine birds. PLoS One [Internet]. 2016 May [cited 2023 Apr 2];11(5): e0154871. Available from: https://doi.org/10.1371/journal.pone.0154871
24. Ratz T, Kremi K, Leissle L, Richardson J, Smiseth PT. Access to resources shapes sex differences between caring parents. Front Ecol Evol [Internet]. 2021 Jul [cited 2023 Apr 2];9:712425. Available from: https://doi.org/10.3389/fevo.2021.712425

25. Thünken T, Bakker TC, Baldauf SA, Kullmann H. Active inbreeding in a cichlid fish and its adaptive significance. Curr Biol [Internet]. 2007 Feb [cited 2023 Apr 3];17(3):225–229. Available from: https://doi.org/10.1016/j.cub.2006.11.053
26. Kokko H, Ots I. When not to avoid inbreeding. Evolution [Internet]. 2006 Mar [cited 2023 Apr 3];60(3):467–475. Available from: http://www.jstor.org/stable/4095309.
27. Cohen LB, Dearborn DC. Great frigatebirds, *Fregata minor*, choose mates that are genetically similar. Anim Behav [Internet]. 2004 Nov [cited 2023 Mar 29];68(5):1229–1236. Available from: https://doi.org/10.1016/j.anbehav.2003.12.021.

10

The Best Mate

All nature's creatures join to express nature's purpose. Somewhere in their mounting and mating, rutting and butting is the very secret of nature itself.

—Graham Swift

10.1 Parental investment and sexual selection

Aedes aegyptii mosquitoes (Figure 10.1) are responsible for transmitting many diseases that affect human health, including the viruses that result in Dengue, Zika, and yellow fever. Females put in a lot of time and resources to manufacture their 100 or so eggs but still need to combine them with those of a male before they can produce offspring. It's a bit unfair, but this new combination of genes gives their offspring a distinct advantage when starting their lives. The females enter a mating swarm and copulate with the first suitable male. It would be best to select a larger male because this is an excellent trait to pass on to both your male and female offspring. After all, size matters when it comes to reproduction for both sexes in *Aedes aegypti*. Fortunately, the sperm can be stored in the female's reproductive tract, so getting the sperm from one male is usually enough for her two to three weeks of adult life. After mating, she needs a blood meal that stimulates the release of hormones that encourage egg production (usually getting this from humans), and, after a few days, will deposit her eggs in a stagnant pool of water.

What about males? Males live about half the amount of time that females do but can reproduce at much higher rates. They don't require blood meals (those nasty mosquitoes that wake you at night are all females) and don't even have the mouthparts enabling them to pierce skin; all their nutrition comes from plant sugars. They can mate three or more times with different females before their sperm is depleted and must be replenished. To maximize the number of offspring they produce, they can be less selective about the females with whom they breed. After all, because they have enough sperm to fertilize many females, and this sperm can be replenished relatively quickly, they don't need to be too fussy. So, they can mate with just about any female they come across. If she produces any offspring, great. If she produces many high-quality offspring, even better! But if they've given their genes to a low-quality female, it's no big deal; there are plenty more sperm where that came from. Now, where's that next female?

The Evolution of Sex. Kevin Lee Teather, Oxford University Press. © Kevin Lee Teather (2024).
DOI: 10.1093/9780191994418.003.0010

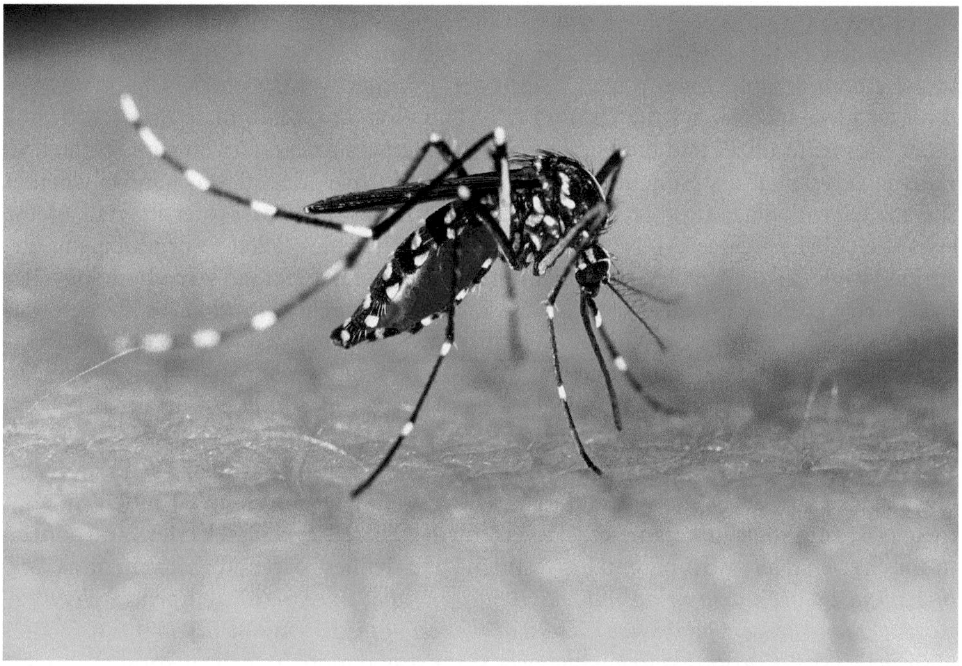

Figure 10.1 *Aedes aegytii mosquitoes nicely illustrate how males and females approach reproduction. Females invest substantially in a small number of expensive eggs and usually mate just once. Males manufacture lots of inexpensive sperm and can mate with many different females.*
Credit: Tacio Philip Sansonovski/Shutterstock Photo ID 369,189,926.

It's not difficult to see how sperm and eggs evolved from gametes of equal size. After all, the first objective of sexual reproduction is to find another gamete with which to combine genetic material, and models indicate that it's easier to do this if one gamete is small and mobile while the other is large, stationary, and sends out signals. Conversely, the implications of this disparity are pretty significant when it comes to maximizing your genetic contribution to the next generation because it already results in an important difference between the sexes. Of course, offspring often require further investment. Usually, this care falls upon the female. Conversely, sometimes males invest substantially in their progeny and are choosier than females about whom they select as mates. However, let's start with the fundamental difference in gamete size between the biological sexes and examine why it may lead to differences in their anatomy and subsequent behaviour.

Eighty years ago, Angus John Bateman realized that the difference in gamete size would affect the reproductive strategies of males and females.[1] Working with fruit flies (*Drosophila melanogaster*) in the early 1940s, Bateman put together five males and five virgin females and monitored their reproductive performance. Because males produce more gametes than females, they have the capacity to contribute to many more progeny. Of course, since all individuals of a sexually reproducing species have one father and one mother, for those males having many offspring there must be males that produce few or no offspring. Less variation in mating success exists for females who invest a considerable amount into fewer eggs. In other words, female *Drosophila*, like those of most species, are the limiting sex. The more females a male mates with, the more eggs he fertilizes and the higher his reproductive success. Not so with females. In Bateman's original experiments, a female's reproductive success was similar whether she mated with one, two, or three males. One male could easily fertilize all of the eggs she could produce. Moreover, only 4% of the females failed to produce any offspring, while 21% of males had no progeny. Although Bateman's methodology has since been criticized,[2] the variability in male and female reproductive success has been verified through many studies of different organisms. Thus, **Bateman's Principle** states that the variation in male fitness is generally higher than that of females. This disparity is particularly true when parents contribute little else to offspring but gametes.

This conclusion was taken a step further by Robert Trivers,[3] who applied Bateman's ideas more extensively, considering that males and females often invest much more in their offspring than just gametes. Trivers further linked sexual selection, introduced in the first chapter, with parental investment. He defined parental investment as *'any investment by the parent in an individual offspring that increases the offspring's chance of surviving (and hence reproductive success) at the cost of the parent's ability to invest in other offspring'*. Trivers proposed that whichever sex invested more in their offspring (and this investment could include the energy invested into gametes, but also things like feeding or guarding the young) became a limiting resource for the sex that invested less. So, the statement by Darwin that 'males compete for choosy females' is usually correct as, most often, the parental investment by females is greater than that by males. But let's be clear: when males invest more than females in their offspring, they may become the limiting sex, and females compete for them. If both males and females invest a substantial amount of time and energy into their progeny, both should be choosy. And this makes sense. Whenever your partner invests heavily in offspring, you'd expect competition for them would intensify. When we look at the strategies of males and females and the classification of mating systems, we will return to Trivers's view on the relationship between parental investment and sexual selection, as the two are intricately connected.

10.2 Intrasexual selection

When the parental investment by one sex exceeds that of the other, individuals of the lower-investing sex compete for mating opportunities. As males are often the sex that invests less, this is often known as male-male competition. However, as I've indicated, sometimes females compete to access males who may provide significant care to the offspring, a situation more fully described in Section 10.4. This competition can be indirect, with no or minimal interactions between the individuals, or direct, involving confrontations.

10.2.1 Indirect competition

One kind of indirect intrasexual selection involves any increase in the ability to detect and reach members of the other sex. Obviously, if you get to a member of the opposite sex faster, you will be in a better position to mate.

After reaching maturity, most male moths forego all other activities, including eating, in order to mate. To make their location known, females often release sex pheromones that males can detect using large feathery antennae (Figure 10.2; often a straightforward way to tell the difference between males and females in this group). For example, about half of the 20,000 sensory hairs on the male silkworm moth (*Bombyx mandarina*) antennae are used to detect bombykol, a pheromone released by the female, and each of these sensory hairs has about 3000 pores through which molecules of bombykol can flow. Because of this high sensitivity, only one molecule of bombykol will result in the neural stimulation of the male antennae,[4] and it takes only a few hundred molecules of bombykol for males to initiate a zig-zag pattern towards the female. The more sensitive his antennae are, the more likely he will find the female and mate.

Male digger bees (there are about 70 species) looking for a mate employ a different strategy. The adult females of these solitary bees dig many nests in the ground, laying one egg in each. After depositing a morsel of food in each, she provides no further care. The egg hatches, the larva feeds on the deposited food, and the young bees come out ready to mate. However, males develop a little more quickly than females; they emerge first and fly close to the ground, waiting for the reproductively receptive females to appear, sensing their location by smell. Several characteristics have been selected in males to make them more successful breeders, including faster development times and increased olfactory capabilities. In a classic study of *Centris pallida*,[5] John Alcock and his colleagues found that males employ two strategies. The first, as described above, is to patrol the area in search of emerging females. The second is to remain stationary near the edge of the emergence site and mate with any virgin female who has not been fertilized by one of the patrollers.

Of course, males may also increase their chance of mating with females if they make themselves more noticeable. This increase in conspicuousness is often done with vocalizations. Advertisement calls of many frogs indicate the location of the male to the female and, of course, his sexual readiness.[6] The louder, more frequently, and longer a frog calls, the more likely females will realize he's (i) of the same species and (ii) looking for a mate. Such advertisement calls also signal a male's location to competing males, reducing territorial intrusions. Of course, this doesn't necessarily mean that you are a good mate—you still have to convince the female of that—but often, the same cues to make yourself most noticeable can be used by females to judge your quality. Often, males change their vocalizations to courtship calls when they detect females nearby.

Of course, a male who's too tired to copulate when a female shows up won't get far in the evolutionary race. Thus, males can increase their reproductive performance by being ready for sex when the opportunity presents itself. Further, it makes sense for a male to continually mate with different females if all he's doing is providing sperm. After

Figure 10.2 *A male gypsy moth has large feathery antennae to detect females. Females remain stationary and release a sex pheromone that stimulates receptors on the male antennae. He then flies to her and copulates for about 5–10 minutes, after which she stops emitting the pheromone and becomes unreceptive to the advances of other males.*

Credit: Hudakore/Dreamstime Photo ID 10,393,249.

all, the new female will likely be more fertile and thus have a better chance of promoting his genes. Males of the red jungle fowl are known to decrease mating with one female while increasing mating with another. So do burying beetles. So do some salamanders. So do a wide range of species studied. There's little reason for a male to keep trying to inseminate a female once he's already had sex with her, particularly if he's providing no further care to offspring. Another female, though? That's different. The propensity to mate with various females has been termed the Coolidge Effect after a joke arising from a visit between a farmer and the former President of the United States. As the story goes, the president and his wife were visiting a farm, and Mrs Coolidge was escorted out to see the chickens. After asking why there was only one rooster, she was told that the rooster could mate many times in one day. 'Tell that to the President', she said. When the president later toured the same area and was given the same information, he inquired, 'Same female every time?' When the farmer replied, 'No, different females', he said, 'Tell that to Mrs. Coolidge.' So, the Coolidge Effect (Figure 10.3) refers to the decline in a male's sexual interest in one female while being sexually aroused by the appearance of a new female.

Although most studies demonstrating the Coolidge Effect have used mammals as study subjects, this phenomenon has been observed in several other groups. For example, male guppies (*Poecilia reticulata*) courted novel females four times longer and put significantly more energy into reproduction than males mating with the same females.[7] Not surprisingly, humans have been subject to many studies to determine if the Coolidge Effect applies to them. As expected, when provided with the option (theoretical only), men sampled across the United States were more (i) likely to spread sex between different females, (ii) excited by their partners continually changing their appearance, and (iii) likely to select novel partners to date.[8] These results suggest that men, more than women, are likely to pursue sexual variety, keeping in line with the Coolidge Effect.

However, before the outcry of male bias, it has recently been shown that females can also experience the Coolidge Effect, although it has only been evaluated in a few studies. Female rats, for example, when provided with the male with which they recently copulated and a new male, prefer the stranger.[9] The effect, however, doesn't appear as strong as in males, nor would we expect it to be. Males gain significantly more than females by repeatedly mating with new females. And although females may benefit by

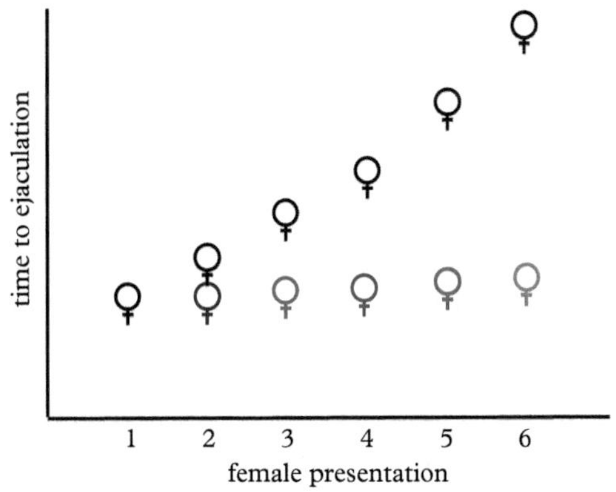

Figure 10.3 *In this hypothetical species (although based on actual data), the male takes progressively longer to ejaculate when mating repeatedly with the same female (indicated by colour). However, when presented with a different female each time, his time to ejaculation is similar for each.*

copulating more than once and with different males, the reasons are varied and discussed more fully in the next chapter.

In addition to being aroused by novel females, males with high endurance levels have a distinct advantage when it comes to mating. Males of the highly dimorphic southern elephant seals (*Mirounga leonina*), discussed before, can be five times as large as females. Large size in males is a distinct advantage as it gives them an edge in direct confrontations with other males while on their mating grounds. However, males fast during the reproductive season. Their food supply is in the water, and they can't afford to leave the breeding grounds on the beach for fear of lost mating opportunities. Thus, those males that come onto the beach in optimal condition are better able to face the energetic demands of a long breeding season.[10]

10.2.2 Direct competition

I've gone through some of the ways in which males can increase their chances of mating with females without having to interact directly with other males. But what if the situation demands they do? Defeating another male in a face-to-face interaction often results in better access to females or increased control over resources necessary for female breeding. Additionally, the contest between males might be used by females to assess which combatants are worthy of copulation. By competing with other males directly, they can also establish themselves in a hierarchy, at least when individuals of the species live in social groups. Let's look at examples of each.

There are many examples of males developing weaponry that they use in contests with other males either to control the resources necessary for successful female breeding or to gain access to the females themselves. These include devices used for pushing, stabbing, lifting, headbutting, fencing, wrestling, or just intimidation. In some cases, females can also use these when selecting a mate. Rhinoceros beetles are a group of about 300 species in which males have one or more rhino-like horns on the anterior parts of their bodies that they use in battles with other males. Not surprisingly, the structural elements of individual horns are closely matched to the fighting style.[11] For example, males of the species *Golofa porteri* have two large horns, the front one on its head oriented towards the back and another on its thorax oriented frontwards. When two males meet on the stalk of a plant being used as a feeding site, a fight often ensues.[12] The larger male, having more substantial horns, flips the competitor unceremoniously off the stalk. In this case, there is no evidence that females prefer males with big horns; the horns simply allow the males to control the resources females need.

The battles between males of all species are an honest signal of strength and fighting ability. Large size in some species carries a clear advantage that can't be faked. Also, large weaponry is generally more effective and is costly to produce; thus, males possessing them must be strong and healthy. So direct confrontations between males represent a straightforward way for females to assess the quality of males, at least as far as strength goes, without devoting much time to evaluation—they can simply watch the interactions and get the sperm from the winner; he should contain the most suitable genes to pass on to her offspring. For example, male black grouse (*Tetrao tetrix*) gather in a display area and compete directly with other males to access the most central locations[13] (Figure 10.4). Intense competition between males often involves displays and vocalizations, although size is an advantage here, and males are 25%–50% larger than females. Females hang around the periphery of display areas, taking in the competition. Males who gain access to preferred regions within the lekking grounds are the ones that are most likely to be chosen by females.

Finally, winning direct interactions in many social species provides males with an elevated status in social hierarchies. The most dominant males typically enjoy preferred access to females when they are fertile. Chacma baboons are dog-faced, social baboons living in Southern Africa's forests and highland grasslands. Each group may have 20–50 individuals, but finding one with over 100 members isn't unusual. The observation that males are about twice as heavy as females suggests that size is also beneficial in this species. Indeed, males will often battle other males to improve their position on a dominance hierarchy, and larger males make better or more intimidating fighters. While physical contests can be brutal and lead to severe injury or death, males often avoid

Figure 10.4 *Two black grouse males during an aggressive interaction at a breeding lek. Large size is instrumental in helping males compete with one another for favoured spots in leks, and in this species, males are significantly larger than females.*

Credit: Mikelane45/Dreamtime Photo ID 34,448,431.

fighting by 'hooting' at other males. These vocalizations are an honest indicator of a male's size and fighting ability, so the victor in male-male contests can often be decided without any physical contact. Dominance has its privileges, as males higher on the social ladder enjoy the company of females much more often than males of lower rank.[14]

10.2.3 Alternative mating strategies

Dung beetles are a type of scarab beetle that feed on, yes, faeces. Some species cut pieces off the dung, make little balls out of it, and roll them to another area where they can be used as a food source for themselves and/or their offspring. Others simply bury the dung where it lies, usually putting little balls of faeces at the end of brood chambers so the offspring has something to eat when emerging from the egg. Despite their diet, some of these dung beetles are quite beautiful. The horned rainbow scarab beetle is a multicoloured dung beetle in which males can possess a prominent horn on their head that they, like rhinoceros beetles, use in battles with other males. In the dung beetle *Onthophagus acuminatus*, males with larger horns (often larger males) defeat other males in direct contests.[15] These males subsequently guard the entrance to brood chambers, where females prepare for the arrival of offspring. However, other smaller males, having small horns that aren't especially useful for fighting dominant males, sneak into the nest, often by digging a new tunnel, and attempt to copulate with the female.

Sneaking is an alternative mating strategy. Suppose a sexually mature male has little chance of winning direct contests with a dominant male, often because of his age or size. He doesn't simply give up. Instead, he may attempt to maximize his genetic contribution by mating with females surreptitiously. Such males hope that any offspring the female has will belong, at least partially, to them. Sneaker or satellite males are known to occur in hundreds of species and can display a range of morphologies and behaviours. Some are younger males who, despite being sexually mature, can't compete with the older, stronger males. These males often change strategies as they age and get larger. Others, like Coho salmon males, may be sneakers all their lives. In this species, sneakers mature earlier, tend to have a higher survival rate, and produce higher-quality sperm than the more common and dominant hooknose males.[16] Side-blotch lizard (*Uta stansburiana*) males actually have three morphs, all applying different mating strategies.[17] Males with orange throats defend large territories and mate with many females; those with blue throats have smaller territories but defend a single female. Those with yellow throats avoid interacting with other males, probably resembling females. The reproductive success of each is dependent on the relative abundance of the other groups; the rare morphs in a population are the most successful and increase in number.

10.2.4 Sperm competition

Competition between males doesn't end when they release their sperm. The sperm cells themselves may have to compete for the right to fertilize the egg. This interaction was probably the first kind of male-male competition after anisogamy had evolved. Sea urchin males and females produce clouds of gametes that float on the water's surface, and sperm from many males compete. If you are a male sea urchin, how can you increase your chance of being a successful breeder? Produce lots of fast-swimming sperm that are easily recognized by the egg once they arrive. Remember, female eggs are faced not only with sperm from many males of their own species but also those of males from other species. The egg recognizes the sperm with proteins on its surface (gamete-recognizing proteins (GRPs)); only males of her own species with the appropriate recognition proteins can get past the egg's defences. Sperm that can be recognized faster will be better positioned to fertilize the egg.

Females that keep eggs inside themselves after being fertilized often copulate with more than one male; in these cases, sperm from different males can also compete for the right to fertilize the egg. Although females may partially control access, males can do many things to maximize their own chances or decrease the chances of other males. First, they can try to prevent sperm competition altogether by restricting the access of other males' sperm to the egg or female's reproductive tract. In many species, such as the house mouse (*Mus musculus*), the male inserts a copulatory plug into the vagina of the female to prevent other males from successfully breeding. Male dunnocks (*Prunella modularis*), a small garden bird found in Europe, peck at the external reproductive bits of females until they eject the sperm of another male. In some species of damselflies, males remove much of the previously stored sperm during copulation with their copulatory organs.

Suppose other males can't be prevented from breeding, and sperm from different males can't be prevented from entering or being removed from the female reproductive tract. In that case, sperm will have to compete directly. Many studies that compare males of the same and different species who are predicted to vary in their level of sperm competition support the prediction that males should produce more sperm when there is a chance that females already have mated or are likely to mate again. What else can they do? We also know that there are various kinds of sperm, each presumably doing different jobs. Many butterflies, moths, molluscs, and fish produce at least two types of sperm—one capable of fertilizing eggs, the other not. The 'Kamikaze hypothesis', developed by evolutionary biologists Robin Baker and Mark Bellis,[18] states that many of the sperm designed by animals (including humans!) remain in strategic places in the female's reproductive tract, trying to intercept the sperm of other males. We know there are differently shaped sperm in one ejaculate, but whether nonfertilizing sperm have distinct functions or are just malformed isn't well-established.

Whatever the case, sperm competition is a no-lose situation for the female, whether fertilization is external or internal. Of the thousands or perhaps millions of sperm that she has received from two (or more) males, the highest-quality individual is expected to have the best chance of producing the one sperm cell that is strong and fast enough to combine its genetic material with that of the female's gamete.

10.3 Intersexual selection

Males may use various anatomical features in indirect or direct competition with other males to increase their breeding chances. The most obvious ones are those that can maximize their fighting ability and, besides general body size, can include structures such as horns, antlers, tusks, canines, and mandibles. The benefit of having such features (as we've seen) is easy to understand—they increase the likelihood of winning direct confrontations with other males and breeding with females. But what about characteristics such as bright colours, long tails, complex vocalizations, and a range of behaviours that would be useless in confrontations with other males? These are directed at females who might use such characteristics to assess potential breeding partners and distinguish between them. But such features, be they anatomical or behavioural, must provide the female with information concerning the quality of the male.

The tail of the male widowbird (*Euplectes progne*) is often used as an example of a trait essential to females in their selection of a mate (Figure 10.5). A Swedish investigator, Malte Andersson, was one of the first to show experimentally that females prefer males with longer tails.[19] Long-tailed widowbirds look something like North American red-winged blackbirds, except, of course, they have long tails. And like redwings, they nest in marshes, and their territory often contains the nests of more than one female with whom they mate. Andersson found that if he decreased the tail lengths of some males by a few centimetres and increased the tail lengths of others by the same amount, females would more likely choose the males with longer tails. But why? What information did females uncover about the males by their tails, and how did this choice evolve?

10.3.1 Runaway selection

Ronald Fisher was one of the founding fathers of population genetics, a biostatistician, and, unfortunately, a staunch supporter of the eugenics movement. That aside, he was the first to describe, in 1930, a mechanism that could explain the evolution of elaborate ornaments of males that were used in mate choice by females.[20] Fisher's argument goes like this. Suppose a male bird had a characteristic, say, a slightly longer tail, which didn't necessarily improve his survival but was preferred by females. Any female who mated with such a male would pass on slightly longer tails, assuming it was a genetically transmissible trait, to her male offspring. She would also pass on, again assuming some genetic basis, the preference for mating with males with longer tails to her daughters. In the following generations, females would be more likely to select males in the population having longer tails. Of course, when they become extreme, having showy plumage or bright colour patterns will likely hinder the male's survival. In other words, even though males might have shorter lives, females would choose them as mates more frequently. Tails would get longer until the advantages of obtaining increased breeding opportunities were overcome by decreased survival.

Figure 10.5 *The male long-tailed widowbird has an impressive tail that is used to convince females to mate with him. Experimental evidence demonstrates that females are less attracted to males with shortened tails and more to males with lengthened tails. But what information is provided by tail length?* Credit: Simon-g/Shutterstock Photo ID 89,657,941.

Remember, the choice by females can be completely arbitrary. Breeding benefits will still be gained by males having such characteristics, and as long as reproductive benefits outweigh the cost of decreased survivorship, the trait will continue to grow. A positive feedback mechanism drives the process: males with the most elaborate ornaments, attractive vocalizations, or brightest colours would have the highest fitness in any breeding generation. Although the decision to mate with such individuals initially may have been arbitrary, the current value of selecting males with such characteristics is related to the increased fitness that accompanies choosing them.

10.3.2 The handicap principle and good genes

Amotz Zahavi was a professor at Tel Aviv University who is well-known for various hypotheses, especially concerning honest signalling. His proposals were often met with scepticism initially, and it wasn't until later that they became the subject of more serious debate. Such was the case for his handicap principle concerning why females preferred males with large, showy ornaments.[21] Zahavi proposed that such traits were honest signals; males couldn't have them if they didn't have the energy to produce them or the strength to outrun the predators that noticed them. By displaying these ornaments, they were saying, 'If I have this (insert elaborate feature here), I must have excellent survival genes because I'm much more noticeable to predators, and it costs a lot to make and carry this thing around.'

As usual, Zahavi's hypothesis resulted in much discussion and many arguments over the next few decades. Additionally, some of the criticisms were valid.[22] Zahavi felt that by having this anatomical add-on, males were telling females that they could still overcome hardships so that they would be good mates. Males who didn't have the trait (or a reduced form of the characteristic) couldn't survive with them, or lacked the energy to produce them, and wouldn't make good mates. The problem is that, eventually, all males would have the trait, and it wouldn't tell females anything; it would simply be costly and subsequently lost from the population. The system would work only if the trait was variable, such that males with very elaborate tails (e.g.) were healthier than those with moderately elaborate tails. Our final hypothesis incorporates this variability.

10.3.3 Parasite resistance

The handicap principle and good-genes hypotheses indicate that females who select high-quality males based on extravagant characteristics will pass good survival genes on to their offspring. However, with any trait that has a high reproductive or survival value, the variance in that trait should decrease over time. Conversely, we know there is substantial variability in the expression of elaborate characteristics, as evidenced by looking at various birds. Thus, some male cardinals are bright red, while others have duller plumage. The length and number of 'eyes' on tail feathers differ among male peacocks. Similarly, the wattles and combs found in many gallinaceous birds, some vultures, starlings, and jacanas (among others) may be brightly coloured, dull, more rigid, or flaccid. Bilateral trait may also express varying degrees of symmetry, and we know that symmetry is an honest signal of health and development. So, females have the opportunity to select among males based on the quality of these characteristics. But what exactly do such traits indicate?

Hamilton and Zuk[23] suggested that elaborate sexually selected characters were associated with parasite loads, such that males having few parasites were able to develop more intense colours or display more elaborate ornaments (Figure 10.6). Again, parasites! Since parasites were constantly coevolving with their hosts, variability was always maintained. Expressed another way, males with the genetic makeup that allowed them to fight off the current pathogens best would provide the best sperm; females should select them as mates as their offspring would benefit from having these genes. So, females could assess the immune system efficiency of males by examining their elaborate traits or behaviours.

Support for this hypothesis has been inconsistent. The main predictions are the most sexually dimorphic species should have the highest parasite loads, and the most elaborately ornamented males within a species should have the lowest loads. The main problem has been that if no correlation is found between the parasitism load and the degree of sexual dimorphism or the trait size within a species, one could always argue that the wrong parasites were being considered. This problem could be addressed by looking not at the level of parasitism directly but at some measure of the male's ability to fight off these infections. The major histocompatibility complex (MHC) refers to a group of highly variable genes that code for proteins able to recognize foreign molecules; thus, it represents a region of the genome directly associated with an individual's immune system ability. Good evidence suggests that females are influenced in their selection of mates in all major vertebrate groups[24] by the MHC of the male. Thus, females could potentially use sexually selected traits to judge a male's ability to battle illness, again, an excellent quality to pass on to your offspring.

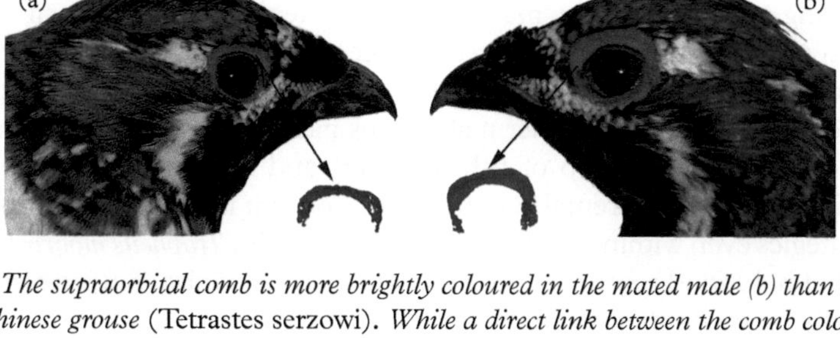

Figure 10.6 *The supraorbital comb is more brightly coloured in the mated male (b) than the unmated male (a) in Chinese grouse* (Tetrastes serzowi). *While a direct link between the comb colour and parasites was not examined, mated males had fewer parasites and higher testosterone than unmated males.*

Credit: CC BY 4.0.

10.3.4 Mitonuclear genes

There is abundant research that indicates sexually selected characteristics are related to the quality of the male. Males that have the most significant ornaments are typically the males that are the healthiest and have the best genes. Females like that. Her offspring will inherit the good genes of her father and thus get a good start in life. While Hamilton and Zuk suggested that females benefited by providing their offspring with good anti-parasite capabilities, a more recent hypothesis has expanded this idea. Earlier, I discussed the interaction between the genes in the mitochondria and the nucleus, mainly when producing an efficient cellular metabolism system. Hill and Johnson[25] suggested that the ability to grow and display an expensive sexually selected trait depends on the mitonuclear genes working well together, resulting in an efficient system to create ATP from oxidative phosphorylation. Thus, by choosing a mate with the best sexually selected trait, she ensures that his nuclear genes will work well with her mitochondrial genes. As the most recent hypothesis to explain the evolution of sexually selected traits, studies are currently evaluating it.

10.3.5 Males can provide more than sperm

It is understood that males can provide something other than good-quality genes. For some of the examples given so far, males increase the chance of their offspring surviving by providing further parental care. This contribution can take several forms. Egg care (e.g. oxygenation, protection, and temperature control), care of newly hatched or born progeny (e.g. feeding, brooding, and protection), and even the care of older offspring (e.g. safety, teaching, and establishment in the social structure) may all be carried out by the male. These activities are valuable to females who may, at least partially, consider them in choosing their mate. However, females have a problem (that may be familiar to humans)—how do they know males will be good parents before mating with them?

In many species, females choose larger males because size matters when protecting females and their offspring. However, female coots (*Fulica americana*) prefer to mate with small, fat males since they can incubate eggs for more extended periods without taking meal breaks.[26] Often, the specific resources that increase the survival of the female or her offspring are controlled by males. Dragonfly females often deposit their eggs on vegetation in shallow water; this requires a place that is well-protected from predators and has a ready food supply for their young after the eggs hatch (Figure 10.7). Males actively defend such sites from other males, thus gaining the opportunity to mate with females. Cichlids, a large group of fish in Africa and South America, contain many species exhibiting diverse parental care patterns. You can find various reproductive and parental strategies even within a cichlid species. In the wild, *Tropheus moorii* males defend territories that hold food resources needed by females. Females, not surprisingly, select males partially based on their territory quality.[27]

In the absence of mate choice based on size or territory quality, males must be able to convince females that they will be good parents. What behaviours might be important? Courtship feeding by males is common in many species where males later feed offspring.

Figure 10.7 *In many animals, one sex (most often the male) defends access to resources that females require for breeding. In these emperor dragonflies* (Anax imperator), *the male (above) is highly territorial around good egg-laying sites for females. The female deposits eggs on vegetation just below the water's surface.*

Credits: male—Perry Van Munster/Adobe Photo ID 448,974,972; female—Chris2766/Adobe Photo ID 61,188,115.

Feeding the female before egg production may provide the female with much-needed nutrition or strengthen the pair bond between male and female. Unfortunately, in most species where courtship feeding has been observed, it only occurs after pair formation, so it is unlikely to be used in mate assessment. In those species where males care for the offspring of various females concurrently, female mate choice may be based, at least partially, on his current parental care abilities. For example, in the glass frog (*Hyalinobatrachium cappellei*), males attend the eggs of various females by keeping them clean and the area free of predators. In this species, a female choice of mate is based not only on the physical qualities of the male and his territory but also on his reliability in caring for eggs.[28] Unfortunately, if and how males convey information about their parental qualities is unknown in most species.

10.4 Do females compete for choosy males?

Most often, the sex that invests more in offspring is tied up with parental duties and is no longer available to mate. Consider mammals. A female is typically receptive to the advances of males for a short period. After their eggs are fertilized, females protect and nourish the developing embryo in their bodies. Once born, the offspring are fed a nutritious substance produced only by its mother. During this time, the female is not available for mating. The males, however, are generally incapable of investing in the offspring after fertilizing the eggs and, since it's easy to produce more sperm, remain in the breeding pool. The number of males and females accessible to the other sex is termed the 'operational sex ratio' (OSR) and is crucial in determining mate selection. Most importantly, individuals available to mate can increase their reproductive output by doing just that—mating. They don't need to be that selective; choosiness takes time, and they are better off using their time giving or getting sperm.

Darwin suggested that males compete for choosy females. This outlook is understandable as, generally, females invest more in progeny and are subsequently available for breeding for a shorter period. However, females aren't always the least available sex. For example, deep-snouted pipefish (*Syngnathus typhle*), like many pipefish species, exhibit sex role reversal. In other words, males invest more heavily in their offspring than do females. In this species, the female does little else but manufacture eggs. After a period of courtship, the female inserts the eggs into a pouch located on the male's abdomen. He fertilizes, protects, and carries them for about five weeks, after which they hatch and swim about freely. Females can produce eggs more rapidly than males can care for a brood. So, in this case, females compete for choosy males.[29] Males must be reasonably confident they're getting high-quality eggs. After all, they don't want to invest heavily in eggs that a low-quality female provides. Similar situations exist in many species. In these situations, the variance in female reproductive output may exceed that of males, a reversal of the trend that we typically find. In many social vertebrates, reproduction is limited to one or a few females. Meerkats (*Surikata suricatta*), for example, live in groups of up to 30 individuals, where one dominant female produces about 80% of the offspring.[30] Subordinate females can breed, but limited resources restrict reproduction to a low level. Such a system invites substantial competition between females for the right to reproduce, even resulting in dominant females killing pups born to subordinates.

There are many other examples of males being the choosier sex or of both males and females being choosy when selecting a breeding partner. Mate choosiness is highly variable between species, as well as within and between sexes. Although there is a substantial focus on female choice, considered responsible for the evolution of many male extravagant features, it is essential to evaluate patterns of parental investment and the OSR in any population. These determine which sex is choosier and which is more likely to compete intrasexually for mates.

10.5 Summary

Males invest little into their gametes, so their success is often determined by how many eggs they can fertilize. They usually compete with other males to access females, or the resources females need to maximize their breeding success. This results in the selection of trait that makes them more successful in such competition. Indirect competition may result in the evolution of characteristics associated with finding and getting to females quickly. Direct interactions between males select for characteristics that often make them better fighters, such as increased size and weaponry. Because a female invests substantially into her gametes, her reproductive success is generally limited by how many eggs she can produce. Females cannot afford to make mistakes, so they tend to be choosy about with whom they combine their genetic material. Often this results in the evolution of male courtship displays, songs, and elaborate ornaments that are useless in male-male interactions but may be useful to her. Such traits reveal something about the male to the female, and although various hypotheses speculate as to what this information might be, none are definitive. Differences in mate-selection strategy are often magnified with added parental contributions by each sex. Generally, females take on more parental duties, although this is variable. Females may have to consider phenotypic traits indicating the male would be a good father after fertilizing the eggs. In other cases, males may select females based on characteristics that suggest they would make good mothers.

References

1. Bateman A. Intra-sexual selection in *Drosophila*. Heredity [Internet]. 1948 Dec [cited 2022 Jun 20];2(3):349–368. Available from: https://doi.org/10.1038/hdy.1948.21
2. Hoquet T. Bateman (1948): Rise and fall of a paradigm? Anim Behav [Internet]. 2020 Jan [cited 2022 Jun 20];164:223–231. Available from: https://doi.org/10.1016/j.anbehav.2019.12.008
3. Trivers RL. Parental investment and sexual selection. In: Campbell B, editor. Sexual selection and the descent of man, 1871–1971. Chicago: Aldine. 1972. pp. 136–179.

4. Syed Z, Kopp A, Kimbrell DA, Leal WS. Bombykol receptors in the silkworm moth and the fruit fly. Proc Nat Acad Sci [Internet]. 2010 May [cited 2022 Jun 20];107(20):9436–9439. Available from: https://doi.org/10.1073/pnas.1003881107
5. Alcock J, Jones CE, Buchmann SL. Male mating strategies in the bee *Centris pallida* Fox (Anthophoridae: Hymenoptera). Amer Nat [Internet]. 1977 Feb [cited 2022 Jun 23];111(977)145–155 Available from: https://doi.org/10.1086/283145
6. Wells K, Schwartz J. The behavioral ecology of anuran communication. In: Narins PM, Feng AS, Fay RR, Popper AN, editors. Hearing and sound communication in amphibians. New York: Springer Handbook of Auditory Research [Internet]. 2007 [cited 2022 Jun 23];28:44–86 (2007). Available from: https://link.springer.com/chapter/10.1007/978-0-387-47796-1_3
7. Jordan LA, Brooks RC. The lifetime costs of increased male reproductive effort: Courtship, copulation and the Coolidge Effect. J Evol Biol [Internet]. 2010 Sep [cited 2023 15 May];23(11):2403–2409. https://doi.org/10.1111/j.1420-9101.2010.02104.x
8. Hughes SM, Aung T, Harrison MA, LaFayette JN, Gallup GG. Experimental evidence for sex differences in sexual variety preferences: Support for the Coolidge effect in humans. Arch Sex Behav [Internet]. 2021 Feb [cited 2023 May 12];50(2):495–509. Available from: https://doi.org/10.1007/s10508-020-01730-x
9. Rosvall KA. Intrasexual competition in females: evidence for sexual selection? Behav Ecol [Internet]. 2011 Dec [cited 2013 May 12];22(6):1131–1140. Available from: https://doi.org/10.1093/beheco/arr106
10. Volzke S, Cleeland JB, Hindell MA, Corney SP, Wotherspoon SJ, McMahon CR. Extreme polygyny results in intersex differences in age-dependent survival of a highly dimorphic marine mammal. R Soc Open Sci [Internet]. 2023 Mar [cited 2023 May 12];10:221635. Available from: http://doi.org/10.1098/rsos.221635
11. McCullough EL, Tobalske BW, Emlen DJ. Structural adaptations to diverse fighting styles in sexually selected weapons. Proc Nat Acad Sci [Internet]. 2014 Sep [cited 2023 May 13];111(40):14484–14488. Available from: https://doi.org/10.1073/pnas.1409585111
12. Eberhard WG. Fighting behavior of male *Golofa porteri* beetles (Scarabeidae: Dynastinae). Psyche [Internet]. 1977 Jan [cited 2023 May 17];84;292–298. Available from: https://repository.si.edu/bitstream/handle/10088/18750/stri_1977_fighting_behavior_of_male_Eberhard_William_G_.pdf?sequence=1&isAllowed=y
13. Alatalo R, Höglund J, Lundberg A. Lekking in the black grouse—A test of male viability. Nature [Internet]. 1991 Jul [cited 2023 May 13];352:155–156. Available from: https://doi.org/10.1038/352155a0
14. Beehner JC, Bergman TJ, Cheney DL, Seyfarth RM, Whitten PL. Testosterone predicts future dominance rank and mating activity among male chacma baboons. Behav Ecol Sociobiol [Internet]. 2005 Oct [cited 2023 May 13];59:469–479. Available from: https://doi.org/10.1007/s00265-005-0071-2
15. Emlen D. Alternative reproductive tactics and male-dimorphism in the horned beetle *Onthophagus acuminatus* (Coleoptera: Scarabaeidae). Behav Ecol Sociobiol [Internet]. 1997 Nov [cited 2023 May 13];41:335–341. Available from: https://doi.org/10.1007/s002650050393
16. Young B, Conti DV, Dean MD. Sneaker 'jack' males outcompete dominant 'hooknose' males under sperm competition in Chinook salmon (*Oncorhynchus tshawytscha*). Ecol Evol [Internet]. 2013 Nov [cited 2023 May 15];3(15):4987–4997. Available from: https://doi.org/10.1002/ece3.869

17. Sinervo B, Lively C. The rock–paper–scissors game and the evolution of alternative male strategies. Nature [Internet]. 1996 Mar [cited 2023 15 May];380:240–243. Available from: https://doi.org/10.1038/380240a0
18. Baker RR, Bellis MA. 'Kamikaze' sperm in mammals? Anim Behav. 1988 Jun;36: 936–939.
19. Andersson M. Female choice selects for extreme tail length in a widowbird. Nature [Internet]. 1982 Oct [cited 2022 Jun 23];299:818–820. Available from: https://doi.org/10.1038/299818a0
20. Fisher RA. The genetical theory of natural selection. Oxford: Clarendon Press; 1930. p. 308.
21. Zahavi A. Mate selection—A selection for a handicap. J Theor Biol [Internet]. 1975 Sep [cited 2022 Jun 23];53(1):205–214.Available from: https://doi.org/10.1016/0022-5193(75)90111-3
22. Penn DJ, Számadó S. The handicap principle: How an erroneous hypothesis became a scientific principle. Biol Rev Camb Phil Soc [Internet]. 2019 Oct [cited 2023 Jun 23];95(1):267–290. Available from: https://doi.org/10.1111/brv.12563
23. Hamilton WD, Zuk M. Heritable true fitness and bright birds: a role for parasites? Science [Internet]. 1982 Oct [cited 2023 Jun 23];218(4570):384–387. Available from: https://doi.org/10.1126/science.7123238
24. E M, Song X, Wang L, Yang Y, Wei X, Gong Y, et al. Mate choice for major histocompatibility complex (MHC) complementarity in the Yellow-rumped Flycatcher (*Ficedula zanthopygia*). Avian Res [Internet]. 2021 May [cited 2022 Jun 23];12:27. Available from: https://doi.org/10.1186/s40657-021-00261-w
25. Hill GE, Johnson JD. The mitonuclear compatibility hypothesis of sexual selection. Proc Biol Sci [Internet]. 2013 Oct [cited 2023 Apr 4];280(1768):20131314. Available from: https://doi.org/10.1098/rspb.2013.1314
26. Petrie M. Female moorhens compete for small fat males. Science [Internet]. 1983 May [cited 2023 May 18];220(4595):413–415. Available from: http://www.jstor.org/stable/1690599
27. Hermann CM, Brudermann V, Zimmermann H, Vollmann J, Sefc KM. Female preferences for male traits and territory characteristics in the cichlid fish *Tropheus moorii*. Hydrobiologia [Internet]. 2015 Apr [cited 2023 May 15];748(1):61–74. Available from: https://doi.org/10.1007/s10750-014-1892-7
28. Valencia-Aguilar A, Zamudio KR, Haddad CF, Bogdanowicz SM, Prado, CP. Show me you care: female mate choice based on egg attendance rather than male or territorial traits. Behav Ecol [Internet]. 2020 May [cited 18 May];31(4):1054–1064 Available from: https://doi.org/10.1093/beheco/araa051
29. Berglund A, Rosenqvist G. Male pipefish prefer dominant over attractive females. Behav Ecol [Internet]. 2001 Jul [cited 2023 Aug 8];12(4):402–406. Available from: https://doi.org/10.1093/beheco/12.4.402
30. Clutton-Brock TH, Brotherton PN, Russell AF, O'Riain MJ, Gaynor D, Kansky R, et al. Cooperation, control, and concession in meerkat groups. Science [Internet]. 2001 Jan [cited 2023 Aug 7];291(5503):478–481. Available from: https://www.science.org/doi/10.1126/science.291.5503.478

11

Not All Sex Results in Reproduction

11.1 Multiple copulations

Smith's longspurs (*Calcarius pictus*) like to copulate (Figure 11.1). On average, they have sex about 365 times for each nest.[1] The mating system has been defined as 'polygynandry', where females usually pair up with two or more males, and males mate with one to three females. After clutch initiation, females copulate at a frequency of more than five times per hour with alpha males and nearly three times per hour with beta males. Interestingly, the females actively solicit males to copulate, and of the solicitations, only a third are successful. If females had their way, longspurs would have sex over a thousand times for each nest! Obviously, this many copulations are unnecessary to fertilize eggs. So why do they engage in sex so often?

Figure 11.1 *Smith's longspurs copulate far more often than needed to fertilize the eggs. As copulations can be expensive, they must serve an evolutionary purpose. The possibilities of repeated copulations are reviewed in this chapter.*

Credit: Female on the left—Paul Roison (USFWS)/CC BY 2.0; Male on the right—Richard Crossley, Sun Jiao/CC BY-SA 3.0.

The Evolution of Sex. Kevin Lee Teather, Oxford University Press. © Kevin Lee Teather (2024).
DOI: 10.1093/9780191994418.003.0011

In many species, individuals have sex repeatedly. This is easy to explain when copulation directly results in more offspring. For example, the polyandrous sandpiper female defends an area containing the territories of more than one male. Those males each have a nest, incubate the eggs, and care for the offspring; they will each mate with the female to produce the eggs in their own nest. In this case, the benefit to the female in copulating more than once is clear: she increases the number of eggs she produces. Similarly, males often increase their reproductive output by copulating with numerous females. However, individuals of both sexes often go beyond engaging in sex the number of times required to fertilize eggs. Copulation overkill is what I want to look at in this chapter.

Both males and females will copulate when there appears to be no advantage to doing so, at least in producing more offspring. However, this does not mean that there is no advantage at all. Note that the goal isn't simply to improve fertilization rates but rather to maximize fitness. In other words, optimizing the condition of one's offspring, in addition to the number, over the individual's entire life is necessary. This may mean breeding with an individual who is not only capable of fertilization but, at least where further parental care is essential, can provide better resources or protection. However, it quickly becomes evident that the reasons for engaging in sexual behaviour differ for males and females; each is pursuing their own strategies for maximizing their overall fitness, not necessarily their current reproductive output.

Table 11.1 provides several reasons why both sexes may copulate repeatedly. Many of these are not mutually exclusive, and it may be difficult to distinguish between them without adequately designed studies. For example, 'Mate assessment' and 'Pair-bond strengthening', two hypotheses proposed for multiple copulations by individuals of a pair, are often difficult to distinguish and have been placed together under **Mate assessment/Ensuring compatibility**. Also, individuals of different species may frequently copulate for different reasons. While evidence exists for each of the hypotheses in Table 11.1, some are more important than others. I'll provide examples for each but let you evaluate them, if you'd like, with further research.

Although this book isn't about humans, our species immediately comes to mind when discussing the role of sexual behaviour outside of reproduction. In our species, such behaviour often has little to do with direct reproduction. Couples consisting of both sexes, or a single sex, may engage in sexual activity for various reasons, and it is an integral part of maintaining a strong bond between individuals; for this reason, sexual behaviour is most often a co-operative venture and is important for both members of the pair (note: I realize that sex can involve more than two individuals, but let's keep it simple). Much has been written on the evolution of mate attraction and sex in humans, including homosexuality, and I direct interested readers to the wealth of information available using a simple web search.

Nearly all of this chapter deals with possible reasons why females mate more than once, with the discussion of males relegated to a short section at the end. The benefit of multiple copulations in males is usually apparent and won't be discussed in detail. However, it may not be as apparent in other cases, particularly when males engage in repeated copulations with one female; thus, the reasons for this are mentioned briefly.

Table 11.1 *Possible reasons for multiple copulations by both sexes.*

	Females	Males
Benefits	1. Bet-hedging 2. Getting materials other than sperm 3. Improving the quality of offspring 4. Mate assessment and ensuring compatibility 5. Ensuring paternity 6. Social reasons	1. Fertilization of multiple females 2. Sperm competition 3. Ensuring your mate is compatible and female coercion 4. Social reasons and play
No benefits or costs	1. Avoiding harassment by other males 2. Forced copulations	1. Misdirected copulation

11.2 Multiple copulations by females

Why do females copulate repeatedly? In most cases, females do so to increase their fitness, directly or indirectly. For example, in a meta-analysis of 122 species of insects, females who mated multiple times had 30%–70% more lifetime offspring production.[2] Fiona Hunter and her colleagues divided repeated copulations by females into those having direct benefits and those used by females to manipulate males.[3] It is also helpful to distinguish between females who mate repeatedly with the same male and females who mate with different males. For example, leaf beetle (*Galerucella birmanica*) females who mate repeatedly with the same male have higher reproductive output than females who mate at the same frequency with different males.[4] However, the opposite is true in seaweed flies (*Coelopa fridgida*); those mating with different males fertilize more eggs.[5] On the other hand, females may not increase their fitness through repeated copulation, especially when conflict between the sexes occurs.

11.2.1 Bet-hedging

Females may mate repeatedly to ensure that they have enough sperm to fertilize their eggs. Repeated copulations are important if males run out of sperm or if the females' mates are sterile; in this case, more than one copulation may be a backup strategy. They may not need the sperm but have sex with several males just in case. For example, female two-spotted field crickets (*Gryllus bimaculatus*) who mate with more than one male have a higher egg-hatching rate, presumably because mating with only one might be risky for getting good quality sperm.[6] Let's look at some other examples.

Copulation only lasts about two minutes in the parasitoid wasp *Spalangia cameroni*. Males of this species are born with all the sperm they will ever have, meaning their supply can run out if they have sex frequently, and they do. Males who copulate between 12 and 52 times exhaust their sperm supplies.[7] Related and well-studied male parasitoid wasps (*Lariophagus distinguendus*) mated with up to 17 females over the course of an hour and continued to copulate with new females even after they ran out of sperm.[8] However, in both these species, females who had mated previously rejected the males, sperm depleted or not. This rejection was not the case in the common lizard, *Zootoca vivipara*. Females who engaged in sex with different males had fewer infertile eggs, suggesting they were avoiding problems with sperm limitation or infertility.[9]

On the other hand, copulation in the firebug *Pyrrhocoris apterus* (Figure 11.2) can be a rather lengthy affair, typically lasting around 12 hours and sometimes up to a

Figure 11.2 *Copulation in the firebug takes a long time, and males will exhaust their supply of sperm after mating with a few females. Thus, females may mate repeatedly to ensure they can fertilize their eggs.*

Credit Bj.schoemakers/CC0 1.0.

week. Firebugs are orange and black insects native to Europe but have more recently been observed in North America and Australia. But don't worry; although often present in large numbers, firebugs are harmless to humans, gardens, and pets. Because copulation takes so long, it's easy to imagine that the sperm of males run out. In fact, it becomes depleted after the fertilization of 3–5 females.[10] Thus, in this species, females must copulate repeatedly to ensure they get enough sperm for their eggs, and in their natural environment, females will mate with a new male, on average, every second day.

Sometimes, a female's mate may be sterile; this would pose a definite problem if producing offspring was your ultimate goal. The function of females soliciting copulations outside of their own relationships in blue tits may be to ensure their eggs are fertilized. In these species, up to 3% of all males are infertile, and such a strategy by females ensures that their own investment in eggs is not wasted.[11] However, fertility insurance does not seem important in zebra finches. Zebra finch pairs often stay together for life, so if a male is infertile, females should be sceptical when the first eggs fail to hatch. However, when the chances of egg hatching are controlled experimentally, females with incomplete hatching clutches were no more likely to leave their mates or seek copulations from other males than when their clutches produced a complete nest of chicks.[12] When they copulate with males other than their social mate, they do so for one (or more) of the reasons listed below rather than possible mate sterility.

11.2.2 Getting materials in addition to sperm

If males provide something to females for the opportunity to copulate, the advantage to females is more apparent. Many male invertebrates provide females with nutritious spermatophores that may increase the females' reproductive performance. Male katydids (*Isophya kraussii*) provide females with spermatophores with enough energy to meet all their nutritional needs for 1 or 2 days, depending on their activities.[13] In the Hawaiian cricket (*Laupala cerasina*), males also present females with spermatophores, but they come in two parts. Initially, **microspermatophores** are provided; these contain no sperm but presumably something of value to females. After females consume these, males deliver **macrospermatophores** that provide sperm.[14] The number of spermatophores accepted by females (and thus the number of copulations) depends on the female's hunger level.

Females could also get more than something to eat when the male transfers sperm to her. Female arctic moths (*Utetheisa ornatrix*), for example, may receive up to 13 spermatophores from different males, which in addition to providing nutrition, deliver a substance that protects her eggs from predation.[15] In Texas field crickets (*Gryllus texensis*), spermatophores contain prostaglandins, an essential lipid compound having numerous beneficial reproductive effects.[16] Prostaglandins are depleted in females during the week after mating, and repeated matings can replenish their levels. Not surprisingly, females who mate multiple times have a higher lifetime fitness than those who copulate only once.

Females of some species may benefit from microbes in the male's semen. There is a reproductive microbiome, distinct from the gastrointestinal microbiome, of which we hear much more, that can positively affect the well-being of males or females. While most of the focus has been on males transferring harmful bacteria during sex, in some cases, females may receive beneficial microbes. Evidence from birds suggests that these microbial communities are similar between monogamous pairs, suggesting that some transfer does occur. However, this is a relatively recent research area, and whether the microbiota contained within semen benefits the female is essentially unknown.[17]

We know that male birds often provide females with food during the nesting period, usually before the eggs are laid, a behaviour known as **courtship feeding** (Figure 11.3). In various species, such as common terns (*Sterna hirundo*), ospreys (*Pandion haliaetus*), red kites (*Milvus milvus*), Montagu's harriers (*Circus pygargus*), lesser kestrels (*Falco naumanni*), and various gull species, courtship feeding is positively correlated with copulation frequency. In some cases, the number or success of copulation attempts is related to the number or quality of prey items, suggesting that the food delivered to females

Figure 11.3 *Common terns are one of many birds that often remain monogamous for more than one breeding season. Males provide females food before pairing and often during the egg-laying period after pairing. Thus, courtship feeding appears to be a way of assessing potential mates and obtaining nutrition while producing eggs.*

Credit: Volodymyr Kucherenko/Adobe Stock photo ID 167,516,764.

influences the copulation number or success rate. In Cooper's hawks (*Accipiter cooperii*), males feed females right after copulation, and females actively solicit copulations from males other than their mates.[18] This suggests they are attempting to (i) maximize their energy intake during egg production or (ii) fertilize their eggs by extrapair males. Male great grey shrikes (*Lanius excubitor*) offer birds, voles, lizards, or insects to females for the right to copulate. What's more interesting is that males offer items of higher nutritional value to females who aren't their mates.[19] It seems that if you are to get females from another pair to mate, you must give them something of value.

Among insects, food-limited females are more likely to engage in copulations in haglids (*Cyphoderris buckelli*), a group related to the crickets and katydids. While copulating, females are permitted to feed on the males' hindwings; this is probably a male's way of ensuring they copulate for long enough to transfer their sperm. In one study, females were separated into two groups, one being food-limited and the other having free access to food.[20] It was predicted that hungrier females would have sex more often, feeding on the delicious hindwings of males. And they did. This experiment provided good evidence that females engaged in extra copulations, not because they required them for fertilization, but to get a needed resource.

Other studies have found that multiple copulations have nothing to do with food. Well-fed female black-legged kittiwakes (*Rissa tridactyla*) are just as likely to copulate with males as are food-deprived females.[21] However, this species exhibits a positive correlation between the rate at which a male provided food and his chance of pairing with that female the following year. Thus, females seemed to be using courtship feeding by males to assess their reliability as a mate. Getting gifts from a male who isn't your mate can also backfire. While female red-winged blackbirds (*Agelaius phoeniceus*) that engaged in sex with neighbouring males were permitted to feed on their neighbour's territories more often, the unfaithful females were often excluded from feeding on their own social male's territory.[22] It appears that there is a price to pay for infidelity.

Often, a male's objective more likely involves discouraging females from repeated copulation to avoid competition with other males. For example, the seminal fluid of the male plant bug (*Lygus hesperus*) contains a chemical that, when taken up by the female, forms a pheromone that renders her unattractive to other males.[23] When the gifts from several male insects are analysed, they often contain substances that reduce the chances of females remating with other males. These are sometimes referred to as **Medea gifts**. In Greek mythology, Medea was the great-granddaughter of the sun god Helios, who assisted Jason in his search for the Golden Fleece, later marrying him. However, after 10 years, Jason tired of her and decided to marry another woman, Aegeus. In anger, Medea sent Aegeus a gift of a poisoned dress and crown; when Aegeus put these on, she burned to death. Nice!

11.2.3 Improving the quality of offspring

It's often difficult to know how females benefit by copulating with males who aren't their mates, especially if they can get enough sperm to fertilize all their eggs from their partner. Perhaps they are trying to improve upon the genes their offspring would have received from their social mate. Maybe the males to whom they're mated were all that remained when all the good males were taken. Or perhaps their mates are good providers, but they fall short in many other characteristics they want to pass on to their sons and daughters. Whatever the reason, studies of many species suggest that females of many socially monogamous species do not limit their copulations to just their partners.

Hide beetles (*Dermestes maculatus*; Figure 11.4) feed on the carcasses of dead animals and are well-known for a few reasons. They have been used to prepare museum specimens by removing decaying tissue from their skeletons. Also, their stage of development can be used in ongoing investigations to determine the time of death of humans. But most importantly to us, females seem to mate with a variety of males to increase the genetic quality of their offspring.[24] Copulating repeatedly with the same male might provide her with some direct benefits, such as care for her offspring, while mating with other males may provide indirect genetic benefits. There is some evidence that, given a choice, females are more likely to remate with different males. However, their mating pattern may also be related to sperm depletion in their mates.

There is substantial evidence that the females of some species mate with males outside their pair bond to reduce risks associated with inbreeding. In red-winged fairy-wrens (*Malurus elegans*), males and females are both philopatric, resulting in close genetic relationships between members of breeding pairs. Presumably, to avoid inbreeding depression in offspring, 70% of the broods contained young fertilized by males outside of their group.[25] However, females do not select extrapair males less

Figure 11.4 *Hide beetles feed on carrion, and their stage of development can provide information about when an animal (including humans) died. When breeding, females of this species often copulate with males other than their mate; this may increase the genetic makeup of their offspring.*
Credit: Jeff Whitlock.

closely related to them than their mate in the noisy miner (*Manorina melanocephala*), where 27% of the broods contain offspring from other males.[26] Thus, while inbreeding avoidance may result in extrapair copulation in some species, it is unimportant in others.

In many species of birds, older males are more successful at extrapair fertilizations. This tendency could be because (i) females favour older males, (ii) older males have an advantage in male-male competition for EPCs, or (iii) older males are more fertile. There is evidence for each of these, depending on the species. The second two reasons are self-explanatory, but why should females favour older males? The reasoning goes, 'If a male has lived long enough, he must have reasonably good genes; these would be useful to pass on to offspring.'. It's nice to know that getting old is a symbol of high genotypic quality!

11.2.4 Mate assessment and ensuring compatibility

Many species of birds, where biparental care is common, mate more frequently than required to fertilize eggs. It's possible that increased copulation rates decrease the time for both males and females to engage in extrapair copulations. Additionally, it may be used to assess mates or strengthen pair bonds. Why use copulations to evaluate potential mates or strengthen pair bonds? Although single acts of copulation are unlikely to be costly for males, multiple copulations increase the expense, especially if (i) males are transferring material to females other than sperm or (ii) it takes time away from other essential activities. In these cases, females could learn about a male's overall condition by seeing if they can copulate repeatedly. In crested tits, for example, females repeatedly solicit potential mates; in over 40% of these solicitations, males turned them down.[27] Interestingly, males judged to be of high quality (determined by their body mass corrected for size) never refused. Thus, the ability to copulate repeatedly could be an honest signal females use to evaluate potential mates.

Ospreys, mentioned previously, are a bird of prey typically found near rivers as they search the water for fish, the main constituent of their diet. They construct a nest, a responsibility of both sexes, at the top of a tree near the water. The female will then lay two or three eggs and incubate them for over a month, although the male may sometimes help. Both parents feed and protect the nestlings after hatching. Since pairs are socially monogamous, often for many years, both partners must be of high quality. And like the longspur mentioned at the beginning of this chapter, ospreys have sex a lot. On average, a pair will copulate just under 300 times over the 45 days prior to having eggs.[28] It's unlikely that so much sex is required to fertilize the eggs, and the observation that many of these copulations take place outside the female's fertile period reinforces the idea that sex isn't being used for reproduction. Newly formed pairs have sex more frequently than well-established pairs, suggesting that individuals of one or both sexes use copulations to assess each other. And since pairs mate before eggs can be fertilized, frequent copulation may also strengthen the ties between the sexes. Males feed females during this pre-fertile period, and the rate of feeding is positively correlated to the rate of copulation; thus, copulation may also be a way of extracting resources from the male by the female.

Similarly, Egyptian vultures (*Neophron percnopterus*), red kites (*Milvus milvus*), American kestrels (*Falco sparverius)*, and merlins (*Falco columbarius*) have all been found to copulate frequently and outside of the period when females can be fertilized. In all cases, strong bonds need to be established by nesting pairs as male help is essential in rearing offspring; in addition to defending the nest, males provide food for their mate, the nestlings, or both. Pairs in these species also usually stay together for more than one breeding season. Evaluating whether your mate (male or female) will make a good partner during the breeding season or strengthening the ties between mating pairs may be the functions of multiple copulations in these species.

11.2.5 Reassuring males of paternity

When fertilization is internal, females are always confident that any resulting offspring came from her, with the possible exception of those that result from brood parasitism. Males can never be sure. Laying eggs or giving birth to live offspring usually comes well after copulation, and males must be a little sceptical about their own biological fatherhood. Of course, whether they only invest sperm or stick around and care for the offspring, males want to make sure they are the ones to fertilize the eggs. Often, they do everything in their power to ensure the offspring belong to them, including increasing their rate of copulation, particularly in species where the last male to provide semen has the best chance of fertilizing the eggs. Females who require that males stick around and help look after the offspring often go to great lengths to assure their partners that they are investing in offspring carrying their genes. Often, that involves copulating with their partners repeatedly.

Dunnocks (*Prunella modularis*) are one of the swingers of the bird world and provide us with some of the best evidence that link repeated copulations and genetic paternity. Dunnocks have been discussed previously and are known to have a variable mating system. Males and females can have one or more partners, and we can find monogamy, polygyny, polyandry, and polygynandry occurring in the same population (see Figure 9.5 and associated text). Since we are focussing on why females mate repeatedly, let's examine polyandry more closely, where females have more than one mate. If the territories of two males overlap with that of a female, she mates with both. The reason? To convince both males that they are the father to at least some of the young in the nest. This persuasion is necessary as the males base their feeding rate on the number of offspring to whom they are genetically related; the more they copulate, the more offspring they father, and the more they feed.[29] Closely related alpine accentors (*Prunella collaris*) have a breeding system similar to that of dunnocks. Like dunnocks, paternal help from two or more males results in greater nesting success. Dominant females copulate more frequently with different males, receive more parental care from males, and have higher breeding success than subordinate females.[30]

In insects for which paternal care is required, assuring males that the offspring belongs to them is also essential. Dung beetles are a large group of insects responsible for the clean-up of faeces, particularly that of large mammals (Figure 11.5). Many species exhibit biparental care, where males assist in rearing offspring. Such care is observed in the dung beetle *Onthophagus taurus*, known as the Taurus scarab, the bull-headed, bull-horned, or horned dung beetle. These beetles have received the title of the strongest insect (although I expect many insects could claim that), being able to pull objects weighing more than a thousand times their weight. They use their super strength to drag faeces underground, where it is used to feed offspring in brood chambers. Males, therefore, can increase the amount of dung available to offspring and thus, the reproductive performance of both males and females, as long as they're the biological fathers. They adjust their contribution based on their confidence in paternity, which depends on how often they copulate with their mate compared with the number of times their mate copulates with other males.[31]

Often, it's difficult to tell if females are copulating repeatedly to strengthen social bonds or to convince their partners that he is the biological father. Perhaps the two strategies have the same goal from the female's point of view—to ensure that the male sticks around and cares for the offspring. For example, the frequent intrapair copulations in any species might assure the males that the resulting offspring are his. This persuading is particularly important if the male continues to provide resources to the offspring. The timing of the copulations may also be a cue. If sex occurs before the possibility of fertilization, it more likely serves as mate assessment or bonding function.

Figure 11.5 *Dung beetles are a diverse group of scarab beetle that bury the dung of other species. The faeces is used as a feed source for the parents and larval beetles when they hatch from eggs, and often, the male assists the female in providing dung for brood chambers. Shown here is a large copper dung beetle.* Credit: berniedup/CC BY-SA 2.0.

11.2.6 Social reasons and pleasure

The first response I get when asking why females should copulate repeatedly is, 'Why not?' The reasoning is that females engage in sex because it is (or can be) an enjoyable experience. This feeling is true for human females, and the more we understand the value of repeated copulation by females of other species, it is likely the case for many of them as well. However, the positive feedback received from a pleasurable experience is a **proximate** response. In examining other species, we must rely on physiological and anatomical similarities that may be similar to those when studying (scientifically) sexual pleasure in humans. For example, eating and drinking when hungry or thirsty provide us (and we assume other animals) with positive feelings, and we know why. An understanding of sexual behaviour is no different; we need to know its evolutionary function if we are to attribute pleasure to it.

Of course, we know that human females have sex for pleasure, even when they don't intend to fertilize eggs. Presumably, sex without reproduction as a goal still serves an evolutionary function. What about other animals? Marc Bekoff, an expert in the social lives of mammals, notes that many mammals seem to enjoy sex, and, without evidence to the contrary, that should be our working hypothesis. We can be relatively confident that males benefit from sex being pleasurable. The more enjoyable it is, the more often they will do it, and the higher will be their reproductive output. If females experience pleasure from sex, we might expect that they would engage in it outside of their fertile period. Indeed, some of them do. Humans, dolphins, and bonobo chimpanzees engage in sexual behaviour throughout the year. The clitoris of female dolphins is similar to humans, fully functional and complete with abundant nerve endings and erectile spaces. In the female bonobo, the clitoris is also large. Not only do females of both species frequently copulate with males, but they also rub clitorises with other females. Of course, sex may not only be pleasurable; it may relieve tensions and provide stronger bonds among individuals in the social group.

11.2.7 Avoidance of harassment from other males

In some cases, females do not benefit from multiple copulations. As discussed in Chapter 8, this often results in a conflict between males and females in the optimum number of copulations for any one reproductive period. In other words, even though females don't benefit, they engage in repeated copulatory acts because males manipulate or force them into it. Because copulation can be costly for females, they may want to do so only at a rate that maximizes their reproductive output. The problem is that the male reproductive potential is much greater, so they usually want to do it as much as possible. This difference may result in the continual harassment of females by males for the opportunity to copulate. The costs to females can either be due to the continued male harassment or the act of reproduction itself. Available evidence suggests that harassment is more costly, and females try to avoid it.

Hermann's tortoises (*Testudo hermanni*) are popular pets, at least for those who think turtles make good pets. Males in most turtles are smaller than females, whose large size corresponds with their egg-laying capabilities. However, the males of this species have a dark side. They can be pretty aggressive to females, biting and often injuring them when trying to get them to copulate. In populations where males significantly outnumber females (i.e. reproductive females are in short supply), up to 75% of the females can be wounded through harassment by males.[32] In New Zealand sea lions (*Phocarctos hookeri*), harassment by males results in the death of 5 in every 1000 breeding females, and 84% of adult females have bite scars from males.[33] These only represent the direct injuries from harassment. Other costs include time, energy, and separation from offspring. In the eastern mosquitofish (G*ambusia holbr*ooki), males harass females for sex, and as a result, female reproductive success declines. This negative effect on fitness occurred even if the males did not copulate—harassed females suffered lower growth and a weaker immune system.[34]

Females can avoid the costs of male harassment in three main ways. First, they can elude the harassing males by simply leaving the area. In this case, they avoid copulating repeatedly. Second, they might be guarded by their own mates—this may involve repeated copulations with him, but presumably, these costs would be lower than those accrued because of the harassment by other males. Third, females can simply give in to other males' advances, thus avoiding the costs of being continually harassed. This 'best of a bad situation' is often called 'convenience polyandry'. There is evidence that each of these plays a role.

Small-spotted catsharks (*Scyliorhinus canicular*) are among the most common types of sharks found along the coast of Europe and North Africa. Females avoid sexually harassing males by taking refuge in shallow water caves after they've been inseminated.[35] Female garter snakes (*Thamnophis sirtalis*) are unlikely to be fertilized by another male after copulating because they receive a mating plug.[36] Of course, this doesn't stop males from trying. Females who have mated, and females who have not, avoid needless harassment by males by dispersing from the area.

Is there evidence that male harassment is reduced by maintaining close ties to your mate? Male pigeons (*Columba livia*) use mate guarding and frequent copulation to protect their paternity.[37] Females usually initiate copulations, especially if their mate defends them from other males. When their partner was removed experimentally, harassment by other males increased considerably, resulting in females feeding less. The authors concluded that females exchanged sex with their partners for increased protection from harassing males. In other species, it may be challenging to determine if females copulate with their mate to assure them of paternity or protect themselves from other males' advances; both may be important. Male stitchbirds (*Notiomystis cincta*) protected their female partners from harassment by other male stitchbirds but not other species, particularly male bellbirds (*Anthornis melanura*); this suggests that males were protecting their paternity investment, not the females.[38]

In some cases, simply giving in to harassing males may be the best strategy for females to avoid the costs associated with their pestering. Males and females of the solitary bee *Anthidium maculosum* (Figure 11.6) are unusual among bees, with males being significantly larger than females. The large size of males presumably helps them defend their territory, which they do for up to three weeks while searching for females. Females

Figure 11.6 *Females in these solitary carder bees travel between territories defended by males to feed on mint plants. To avoid harassment by territory holders, she simply mates with them despite not needing their sperm.*

Credit: Evan Dankowisz/CC BY 4.0.

foraging within these territories often mate with the owner of that territory, even though she gets enough sperm to fertilize her eggs from a single copulation. It may save time and energy and reduce foraging interruptions to copulate passively for 30 seconds rather than trying to evade or repel the large, aggressive males.[39] Female kelp or seaweed flies (*Coelopa frigida*) are constantly harassed by males who will mount a female more than once every 10 minutes. It is estimated that females, who live for three weeks, may experience hundreds of matings in their lives, far more than necessary to fertilize their eggs. Since they don't seem to leave the wrack beds in which they live, they cannot avoid this male harassment and are thought to copulate repeatedly as a form of convenience polyandry.[40]

11.2.8 Forced extrapair copulations

Copulation can be costly for several reasons, and females clearly avoid harassment by males when possible. But sometimes, females are forced to copulate, generally by males other than their mates. We've looked at forced copulation previously as they provide a clear example of sexual conflict. Females are not willing participants in sexual activity but suffer both the costs of copulation and overt aggression by males. Recall that in mallards, a species where males attempt to force females to copulate, up to 10% of females may be killed as a result of male aggression. Do parallels occur in humans? Let me say that there is substantial controversy concerning the underlying causes of forced copulation, or rape, in humans. However, arguments often arise over proximate and ultimate explanations, so be sure to understand the difference!

11.3 Multiple copulations by males

There is little to say about repeated male copulation that hasn't been dealt with previously. Males often engage in sex many times during a reproductive bout, especially with different females, as it usually improves their reproductive output. But why do they mate with the same female numerous times even though (i) she doesn't require much sperm to fertilize her eggs and (ii) many of these copulations occur outside the fertile period? Males may repeatedly mate with the same (or another) female to (i) promote themselves or strengthen the pair bond (discussed in Section 11.2.4), (ii) compete with other males through sperm competition (Section 10.2.4), (iii) elevate themselves socially (similar to females: Section 11.2.6), or (iv) practice (many male mammals engage in 'pseudo-reproductive' behaviour when young, presumably practicing for the real thing). In some cases, males misdirect their sexual energy. For example, I once saw a red-winged blackbird male trying to have sex with a dead robin that had been hit by a car. I don't even know if it was a female robin. This observation illustrates the intense drive for copulation that has been selected in males.

11.4 Summary

Although both sexes often copulate repeatedly during a single reproductive bout, most focus has been on females. Presumably, females can get enough sperm in one ejaculation to fertilize their eggs, so it often isn't clear why she mates repeatedly. However, there are several reasons why females can increase their fitness by mating numerous times with the same or many males. First, the initial male with whom they copulate may be infertile or provide insufficient sperm to fertilize her eggs. Second, she may get resources, in addition to sperm, when she copulates. Third, she may improve the genetic quality of her offspring by mating with a male other than her social partner. Fourth, in species in which biparental care is required to rear offspring successfully, multiple copulations may be a way of strengthening the pair bond or assessing the quality of your mate. Fifth, convincing males of their paternity may be essential, especially in species where males are required to rear genetically related offspring. And lastly, copulation may be used in social species to maintain or enhance their position in the group. In some cases, females may copulate repeatedly even though it negatively affects fitness, especially when there is sexual conflict regarding the copulation rate. This may happen if males harass females for sex or force females to mate.

Multiple male copulations are easier to understand, and the reason is typically straightforward. However, in addition to increasing the number of eggs they fertilize, they may repeatedly mate with the same female for mate assessment or because females coerce them. Sperm competition may be another reason for prolific mating, either with the same female or different females; while increased numbers of sperm may not be necessary for fertilization, repeated copulation increases your mating success when competing with other males.

References

1. Briskie JM. Copulation patterns and sperm competition in the polygynandrous Smith's Longspur. Auk [Internet]. 1992 Jul [cited 2023 May 28];109(3):563–575. Available from: https://doi.org/10.1093/auk/109.3.563
2. Arnqvist G, Nilsson T. The evolution of polyandry: Multiple mating and female fitness in insects. Anim Behav [Internet]. 2000 Aug [cited 2023 May 28];60(2):145–164. Available from: https://doi.org/10.1006/anbe.2000.1446
3. Hunter FM, Petrie M, Otronen M, Birkhead T, Møller AP. Why do females copulate repeatedly with one male? TREE [Internet]. 1993 Jan [cited 2023 May 28]:8(1):21–26. Available from: https://doi.org/10.1016/0169-5347(93)90126-A
4. Wang L, Meng M, Wang Y. Repeated mating with the same male increases female longevity and fecundity in a polyandrous leaf beetle *Galerucella birmanica* (Coleoptera: Chrysomelidae). Physiol. Entomol [Internet]. 2018 Feb [cited 2023 May 28]:43(2):100–107. Available from: https://doi.org/10.1111/phen.12233
5. Dunn DW, Sumner JP, Goulson D. The benefits of multiple mating to female seaweed flies, *Coelopa frigida* (Diptera: Coelpidae). Behav Ecol Sociobiol [Internet]. 2005 Mar [cited 2023 May 28];58(2):128–135. Available from: http://www.jstor.org/stable/25063595
6. Yasui Y, Yamamoto Y. An empirical test of bet-hedging polyandry hypothesis in the field cricket *Gryllus bimaculatus*. J Ethol [Internet]. 2021 May [cited 2023 Sep 6];39:329–342. Available from: https://doi.org/10.1007/s10164-021-00707-0
7. King BH. Sperm depletion and mating behavior in the parasitoid wasp *Spalangia cameroni* (Hymenoptera: Pteromalidae). Great Lakes Entomol [Internet]. 2000. [cited 2022 Nov 16];33(2):117–127. Available at: https://scholar.valpo.edu/tgle/vol33/iss2/4
8. Steiner S, Henrich N, Ruther J. Mating with sperm-depleted males does not increase female mating frequency in the parasitoid *Lariophagus distinguendus*. Entomologia Experimentalis et Applicata, 2008 Nov [cited 2022 Nov 17];126(2):131–137. Available from: https://doi.org/10.1111/j.1570-7458.2007.00641.x
9. Uller T, Olsson M. Multiple copulations in natural populations of lizards: Evidence for the fertility assurance hypothesis. Behaviour [Internet]. 2005 Jan [cited 2022 Nov 17];142(1):45–56. Available at: https://www.jstor.org/stable/pdf/4536228.pdf
10. Honěk AL, Martinková Z, Brabek M. Mating activity of *Pyrrhocoris apterus* (Heteroptera: Pyrrhocoridae) in nature. Eur J Entomol [Internet]. 2019 May [cited 2022 Oct 20];116:187–193. Available from: https://doi:10.14411/eje.2019.020
11. Santema P, Teltscher K, Kempenaers B. Extrapair copulations can insure female blue tits against male infertility. J Avian Biol [Internet]. 2020 Apr [cited 2022 Oct 20];51(6). Available from: https://doi.org/10.1111/jav.02499
12. Ihle M, Kempenaers B, Forstmeier W. Does hatching failure breed infidelity? Behav Ecol [Internet]. 2013 Jan [cited 2023 Oct 21];24(1):119–127 Available from: https://doi.org/10.1093/beheco/ars142

13. Voigt CC, Michener R, Kunz TH. The energetics of trading nuptial gifts for copulations in katydids. Physiol Biochem Zool [Internet]. 2005 May [cited 2023 Jan 6];78(3):417–423. Available from: https://doi.org/10.1086/430224
14. Shaw KL, Khine AH. Courtship behavior in the Hawaiian cricket *Laupala cerasina*: Males provide spermless spermatophores as nuptial gifts. Ethology [Internet]. 2004 Feb [cited 2023 Jan 6];110(2):81–95. Available from: https://doi.org/10.1046/j.1439-0310.2003.00946.x
15. Lamunyon C. Increased fecundity, as a function of multiple mating, in an arctiid moth, *Utetheisa ornatrix*. Ecol Ent [Internet]. 1997 Feb [cited 2023 Jan 6];22(1):69–73. Available from: https://doi.org/10.1046/j.1365-2311.1997.00033.x
16. Worthington AM, Jurenka RA, Kell CD. Mating for male-derived prostaglandin: A functional explanation for the increased fecundity of mated female crickets? J Exp Biol [Internet]. 2015 Sep [cited 2023 Jan 6];218(17):2720–2727. Available from: https://doi.org/10.1242/jeb.121327
17. Ma ZS. Microbiome transmission during sexual intercourse appears stochastic and supports the Red Queen hypothesis. Front Microbiol [Internet]. 2022 Mar [cited 2023 Sep 6];12:789983. Available from: doi:10.3389/fmicb.2021.789983
18. Rosenfield RN, Sonsthagen SA, Stout WE, Talbot SL. High frequency of extrapair paternity in an urban population of Cooper's Hawks. J Field Ornithol [Internet]. 2015 May [cited 2022 Dec 14];86(2):144–152. Available from: https://doi.org/10.1111/jofo.12097
19. Tryjanowski P, Hromada, M. Do males of the great grey shrike, *Lanius excubitor*, trade food for extrapair copulations? Anim Behav [Internet]. 2005 Mar [cited 2022 Dec 16];69:529–533. Available from: https://doi.org/10.1016/j.anbehav.2004.06.009
20. Judge KA, De Luca PA, Morris GK. Food limitation causes female haglids to mate more often. Can J Zool [Internet]. 2011 Oct [cited 2022 Dec 17];89(10):992–998. Available from: https://doi.org/10.1139/z11-078
21. Helfenstein F, Wagner RH, Danchin E, Rossi J-M. Functions of courtship feeding in black-legged kittiwakes: Natural and sexual selection. Anim Behav [Internet]. 2003 May [cited 2022 Dec 16];65(5):1027–1033. Available from: https://doi:10.1006/anbe.2003.2129
22. Gray, EM. Female red-winged blackbirds accrue material benefits from copulating with extra-pair males. Anim Behav [Internet]. 1997 Mar [cited 2022 Dec 15];53(3):625–639. Available from: https://doi.org/10.1006/anbe.1996.0336
23. Brent CS, Byers JA, Levi-Zada A. An insect anti-antiaphrodisiac. eLife [Internet]. 2017 Jul [cited 2023 Jun 2];6:e24,063. Available from: https://doi.org/10.7554/eLife.24063
24. Archer MS, Elgar MA. Female preference for multiple partners: Sperm competition in the hide beetle, *Dermestes maculatus* (DeGeer). Anim Behav [Internet]. 1999 Sep [cited 2023 Jan 7];58(3):669–675. Available from: https://doi.org/10.1006/anbe.1999.1172
25. Brouwer L, Van De Pol M, Atema E, Cockburn A. Strategic promiscuity helps avoid inbreeding at multiple levels in a cooperative breeder where both sexes are philopatric. Mol Ecol [Internet]. 2011 Nov [cited 2023 Jun 2];20(22):4796–4807. Available from: https://doi.org/10.1111/j.1365-294X.2011.05325.x
26. Barati B, Andrew RL, Gorrell JC, McDonald JC. Extrapair paternity is not driven by inbreeding avoidance and does not affect provisioning rates in a cooperatively breeding bird, the noisy miner (*Manorina melanocephala*). Behav Ecol [Internet]. 2018 Jan [cited 2023 Jun 2];29(1):244–252. Available from: https://doi.org/10.1093/beheco/arx158
27. Lens L, Van Dongen S, Van den Broeck M, Van Broeckhoven C, Dhondt AA. Why female crested tits copulate repeatedly with the same partner: Evidence for the mate assessment hypothesis. Behav Ecol [Internet]. 1997 Jan [cited 2023 Dec 15];8(1):87–91. Available from: https://doi.org/10.1093/beheco/8.1.87

28. Mougeot F, Bretagnolle V, Thibault J-C. Effects of territorial intrusions, courtship feedings and mate fidelity on the copulation behaviour of the osprey. Anim Behav [Internet]. 2002 Nov [cited 2022 Dec 14]; 64(5):759–769. Available from: https://doi.org/10.1006/anbe.2002.1968
29. Davies NB, Hatchwell BJ, Robson T, Burke T. Paternity and parental effort in dunnocks *Prunella modularis*: How good are male chick-feeding rules? Anim Behav [Internet]. 1992 May [cited 2023 Apr 3];43(5):729–745. Available from: https://doi.org/10.1016/S0003-3472(05)80197-6
30. Nakamura M. Multiple mating and cooperative breeding in polygynandrous alpine accentors. I. Competition among females. Anim Behav [Internet]. 1998 Feb [cited 2023 Apr 3];55(2):259–275. Available from: https://doi.org/10.1006/anbe.1997.0725
31. Hunt J, Simmons LW. Confidence of paternity and paternal care: Covariation revealed through the experimental manipulation of the mating system in the beetle *Onthophagus taurus*. J Evol Biol [Internet]. 2002 Aug [cited 2023 Apr 3);15(5):784–795. Available from: https://doi.org/10.1046/j.1420-9101.2002.00442.x
32. Golubović A, Arsovski D, Tomović L, Bonnet X. Is sexual brutality maladaptive under high population density? Biol J Linn Soc [Internet]. 2018 Jul [cited 2023 May 12];124(3):394–402. Available from: https://doi.org/10.1093/biolinnean/bly057
33. Chilvers BL, Robertson BC, Wilkinson IS, Duignan PJ, Gemmell NJ. Male harassment of female New Zealand sea lions, *Phocarctos hookeri*: Mortality, injury, and harassment avoidance. Can J Zool [Internet]. 2005 May [cited 2023 May 12];83(5):642–648. Available from: https://doi.org/10.1139/z05-048
34. Pilastro A, Benetton S, Bisazza A. Female aggregation and male competition reduce costs of sexual harassment in the mosquitofish *Gambusia holbrooki*. Anim Behav [Internet]. 2003 Jun [cited 2023 May 12];65(6):1161–1167. Available from: https://doi.org/10.1006/anbe.2003.2118
35. Wearmouth VJ, Southall EJ, Morritt D, Thompson RC, Cuthill, IC, Partridge JC, et al. Year-round sexual harassment as a behavioral mediator of vertebrate population dynamics. Ecol Mono [Internet]. 2012 Aug [cited 2023 May 13];82(3):351–366. Available from: https://doi.org/10.1890/11-2052.1
36. Friesen CR, Shine R, Krohmer RW, Mason RT. Not just a chastity belt: The functional significance of mating plugs in garter snakes, revisited. Biol J Linn Soc [Internet]. 2013 Jul [cited 2023 May 14];109(4):893–907. Available from: https://doi.org/10.1111/bij.12089
37. Lovell-Mansbridge C, Birkhead TR. Do female pigeons trade pair copulations for protection? Anim Behav [Internet]. 1998 Jul [cited 2023 14 May];56(1):235–241. Available from: https://doi.org/10.1006/anbe.1998.0774
38. Low M. Factors influencing mate guarding and territory defence in the stitchbird (hihi) *Notiomystis cincta*. New Zeal J Ecol [Internet]. 2005 [cited 2023 May 14];29(2):231–242. Available from: http://www.jstor.org/stable/24058179
39. Alcock J, Eickwort GC, Eickwort KR. The reproductive behavior of *Anthidium maculosum* (Hymenoptera: Megachilidae) and the evolutionary significance of multiple copulations by females. Behav Ecol Sociobiol [Internet]. 1977 Dec [cited 2023 May 16];2(4):385–396. Available from: http://www.jstor.org/stable/4599147
40. Blyth JE, Gilburn AS. Extreme promiscuity in a mating system dominated by sexual conflict. J Insect Behav [Internet]. 2006 Jul [cited 2023 May 16];19:447–455. Available from: https://doi.org/10.1007/s10905-006-9034-3

12

Mating Systems

12.1 Problems in classifying mating systems

People love to classify things. It makes them easier to study and understand. This grouping of topics is always brought home in a university course—understanding concepts or remembering facts is much easier if they are lumped together under individual headings. Quite often, subjects logically fit into categories; other times, they are more forced. Mating systems belong to the latter. There have been two main problems. First, until recently, we have focussed primarily on what males were doing. While it is understandable to think that this approach resulted from a chauvinistic slant to science in general, this is not entirely correct. Males tend to be more visible in many species, with conspicuous courtship rituals and noticeable physical features. Females are often more secretive and go about their sexual business without much fanfare.

Second, populations more often contain a variety of mating systems. Each individual, male or female, faces unique circumstances they must evaluate to maximize their fitness. For example, polygyny is often regarded as one male having a sexual relationship with more than one female, while the females are assumed to copulate only with that male. Most likely, though, **some** males may be polygynous, while others are monogamous, and many don't mate at all. Females in the same population may mate with one or more males, depending on opportunity and needs. The diversity of possible outcomes makes it difficult to label a population, let alone a species, as having a specific mating system. In particular, assigning parentage without modern molecular techniques is often difficult. Since many investigations address evolutionary questions concerning reproductive strategies, information about the genetic parents is crucial.

While it's essential to examine the best options for breeding success faced by both males and females when they reproduce in a given environment, many generalizations can still be made. For example, certain environmental conditions allow males to defend a group of females, while others make defending the resources required by breeding females easier. Defending anything is impossible in other cases, and males assume a different reproductive strategy. Females, on the other hand, may get sufficient sperm from one high-quality male that they can use to fertilize all their eggs during that reproductive bout. In many cases, however, as we have seen in Chapter 11, a female often benefits by

The Evolution of Sex. Kevin Lee Teather, Oxford University Press. © Kevin Lee Teather (2024).
DOI: 10.1093/9780191994418.003.0012

Table 12.1 *Costs and benefits of an individual or its mate copulating with more than one individual.*

	Benefits of copulating with > 1 individual	**Costs of copulating with > 1 individual**	**Costs of mate copulating with > 1 individual**
Females	Insurance Good genes More parental care Increased reproductive output	Detection by mate Increased exposure to predators and parasites	Reduced resources Reduced parental care Sperm depletion (?)
Males	Increased reproductive output	Detection by mate Increased exposure to predators and parasites	Cuckoldry Decreased reproductive output

mating with more than one male. In some instances, particularly when parental care is required, a female may mate with separate males, providing them each with offspring to look after and reducing her own level of care.

So, although I use the general terms 'monogamy' (i.e. one male, one female), 'polygyny' (i.e. one male, more than one female), and 'polyandry' (i.e. one female, more than one male), it is necessary to stress that any patterns we find are based on the costs and benefits of various options for both sexes (Table 12.1). For example, a male can increase his mating success in the longnose filefish (*Oxymonacanthus longirostris*) by breeding with more than one female (i.e. polygynously).[1] In contrast, females could improve their reproductive success if their partner only bred with them (i.e. monogamously). Although this species has no parental care, females who mate monogamously produce more eggs. Since both parents are rarely needed to rear offspring, mating behaviour is more correctly viewed in terms of sexual conflict rather than a co-operative venture between males and females. Thus, as we go through the possible mating systems, remind yourselves that how any individual reproduces is determined by its own condition and the circumstances faced when breeding.

12.2 What factors promote monogamy?

Are there any truly (genetically) monogamous species where one male and one female produce offspring together? Of course, although it is relatively rare because one of the sexes can usually increase their reproductive output by mating with more than one individual. In many cases, some individuals in a population are forced to remain monogamous because of their mates' behaviour.

Schistosomiasis is a common tropical disease that affects well over 200 million people and is generally found in regions where conditions, especially water, are unsanitary. A fluke, a form of flatworm, causes the disease and the symptoms brought on by the body's reaction to its eggs that can develop in the digestive or urinary tract. Depending on which system is infected, symptoms can include diarrhoea, blood in the stool, abdominal pain, liver enlargement, blood in the urine, bladder or kidney damage, and various problems associated with the reproductive tracts of both men and women. Although the symptoms are chronic, it is hard to get an accurate count of the number of deaths that arise from the disease as the actual cause of mortality may be attributed to another problem. *Schistosoma mansoni* is the scientific name of one of the flukes (of five or six) that cause the disease. The sexes are separate and genetically determined at the time eggs are fertilized. After about three weeks of development, involving some time in snails, male and female flukes enter their definitive hosts (humans), where breeding occurs. Interestingly, females will not fully mature if they don't come in contact with a male, although males do fine without females. When a male and female fluke meet, the latter enters a groove on the male's body and will generally remain there for the rest of her life. She produces 200–300 eggs daily until she dies a few weeks later. Nearly all of the evidence suggests that these flukes remain monogamous.[2] However, there are a few reports of mate switching and even two females inhabiting the body groove of males at once, although it's possible that the male mates with only one.

In flukes, no further parental care is invested in offspring once the eggs are generated. Since females produce eggs every day after they mature and live about the same amount of time as males, it is difficult for a male to improve his reproductive performance by kicking one female out of his groove and looking for another. Indeed, he may be taking a considerable risk by doing so. Females are in the same position. Presumably, she's chosen him for his genetic quality, as larger males are preferred because of their greater reproductive success. On the other hand, males may be in short supply, so she enters the groove with the first one she comes in contact with. In either case, it would be in her best interests to remain with him and breed. So, monogamy works out well for male and female flukes.

Monogamy will result when the benefits of staying with one mate exceed its costs for both males and females. This situation can occur for many reasons. First, spatial patterns may be such that encountering another reproductively available mate is unlikely. Second, males must stay around and defend their genetic investment, or females are aggressive to other females if sharing them with their mates results in decreased resources or care. In either case, one sex limits breeding opportunities for the other. Third, both parents require parental care to rear their offspring successfully, and there is little opportunity to mate with other individuals. Of course, these conditions only encourage monogamous relationships. Indeed, male and female aggression to members of their own sex usually indicates that copulating outside of a monogamous relationship is at least a threat, if not common. Also, more than one of these reasons can result in monogamous relationships in a given population.

Let's start with the most straightforward situation—males and females are expected to mate monogamously if the chance of finding another individual of the opposite sex in breeding condition is very low. Caribbean cleaning or sharknose gobies (*Elacatinus evelynae*; Figure 12.1) are small fish having the odd habit of cleaning (and eating) the

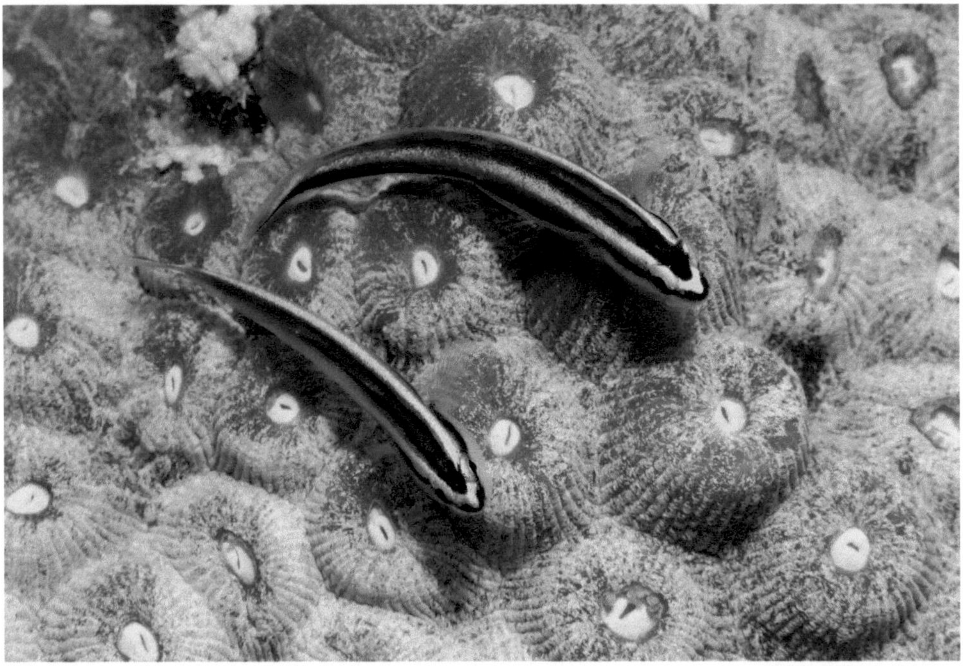

Figure 12.1 *A pair of sharknose gobies swim together on a coral head. Most males and females of this species are monogamous, possibly because it would take too long, or is too dangerous, to find another mate.*

Credit: gonepaddling/123RF ID: 98,487,165.

parasites off the skin of much larger species. In this mutualistic activity, the goby gets a meal, and the larger fish gets its skin freed of potentially nasty organisms. For these gobies, a large breeding partner benefits both males and females—large females have more eggs that can be fertilized, and large males can better protect females while they're foraging. Generally, once males and females get together, they stay together, at least for several reproductive periods. At least part of the reason that Caribbean cleaning gobies are monogamous is that it would take too long to find another mate, especially one larger than their present one.[3] This isn't to say that they wouldn't if the opportunity arose. Both males and females will act aggressively to any member of their own sex that gets too close.

If the females are widely dispersed, the best strategy for a male might be to stay with one female, protecting her and perhaps helping her raise the offspring if his assistance increases the success rate. This situation may be why a few species of mammals are monogamous. In most mammals, males repeatedly mate with as many females as possible, which is unsurprising because females must provide early care for their offspring (gestation and lactation), while males can contribute little during this period. However, it is more difficult for males to have more than one mate when females are widely dispersed because resources are spread out or because females are aggressive towards other females.[4] In such cases, they remain monogamous.

12.2.2 Mate-enforced

Sometimes, both sexes are forced into monogamy because males will remain and guard their mates against the advances of other males. Typically, this explains social monogamy, but males and females may sneak off and copulate with another individual when their mate isn't looking. For example, once threatened with extinction, Seychelles warblers (*Acrocephalus sechellensis*) live on only a few islands. Females produce a single egg during a short breeding season, and once a male and a female establish a territory, they usually remain together until one dies. However, this presents a real conflict. Males can double their reproductive performance if they fertilize the neighbouring female (assuming they fathered the offspring on their territory). Thus, males often guard their mates against other males' advances to ensure their paternity, watching over them when the egg can be fertilized while leaving the females to their own devices when she is not fertile. Despite this, the frequency of extrapair paternity is high, and more than 40% of eggs laid don't belong to the presumed father. Thus, while many aren't genetically monogamous, extrapair mating may be limited due to the behaviour of their mates.

The mating system of Seychelles warblers suggests that males can restrict extrapair copulations by guarding their mates and hindering their liaisons with other males. However, sometimes, the behaviour of females reduces the chances of their mate breeding with someone else by acting aggressively towards any other females that come around. Why do females do this? After all, sperm is plentiful, and there is usually enough for everybody. However, if the female requires something else that the male must provide, either resources or care for her offspring, and his mating with another female jeopardizes that provision, we expect her to act accordingly. Razorbills or lesser auks (*Alca torda*) are sexually monomorphic, socially monogamous, colonial seabirds. Pairs only start breeding after three years of age and will generally mate for life, but both sexes may try to better their reproductive performance by mating with other individuals. Males guard females against the advances of other males and are, for the most part, successful. Female aggression is more interesting. Males not only help incubate the eggs and feed the nestlings, but also accompany the offspring to the sea and continue to protect and feed them for a few more weeks. So, females have much to lose if their mate leaves for another razorbill. This situation isn't likely to happen if the other female already has a mate, but some unmated females may actively solicit copulations from mated males.[5] Probably for this reason, females will vigorously and aggressively defend males from the advances of potential rivals.

12.2.3 Parental care

In some cases, organisms invest little in their offspring, and males and females are free to mate with other individuals if advantageous to them. At the other extreme are species in which caring by two parents is necessary to maximize the reproductive potential of both sexes. Many bird species are socially monogamous, but closer examination reveals that males and females may both cheat on their mates. However, this genetic unfaithfulness doesn't seem to be the case for some. For example, black vultures (*Coragyps atratus*) must invest substantially in their offspring. Although no extensive nest is built, females lay two eggs, and both parents will incubate them for about five weeks. After this, the father and mother will take turns regurgitating food for them that they usually get from carcasses, often for another eight weeks. There is no evidence that either sex will copulate with any vulture other than their mate.[6] Perhaps living in social groups, often with relatives, makes the costs of being unfaithful too high. Or maybe they simply don't have time.

We often hear about lifetime monogamy in some birds. However, despite seeing it in a few species, such as wandering albatrosses (*Diomedea exulans*; Figure 12.2), bald eagles (*Haliaeetus leucocephalus*), and black vultures, it is relatively rare. What do individuals gain? Wouldn't one or both sexes be better off if they sought out other mates with whom to combine their genetic material? Besides having a significant level of parental care, the courtship period in such species is extensive. It takes a lot of time to assess the potential

Figure 12.2 *Sometimes courtship takes a long time, often many years, before pairs mate. This time commitment probably restricts most individuals to one mate unless their mate dies when relatively young. These wandering albatrosses don't breed until they are at least 10 years old, and a male and female may be together for years before mating.*
Credit Ed Dunens/CC BY-2.0.

value of a mate. Because the care contributed by one's partner is generally essential to raising offspring, any mistakes in selecting a mate can be disastrous. For example, wandering albatrosses aren't fully mature until they're about 10 years old, and pair formation can take many years. Plus, after a long winter apart, they engage in extensive courtship rituals when they return to the breeding ground. So, monogamy might result from time limitation. With this lengthy period of mate assessment, individuals would take a big chance in engaging in extrapair matings or changing mates and probably have little opportunity to do so.

12.3 We would expect polygyny to be common

The number of offspring a female produces, or her reproductive output, is not usually nearly as variable as it is for males. A simple example can be used to illustrate this. Let's say a female produces five offspring, and both sexes can copulate with up to three partners. The potential number of offspring a male can have is 0 to 15, while a female's reproductive output is 0 to 5. For a female, there may be some difference in the genetic quality of her offspring but generally not the actual number, given that her partner can usually fertilize all of her eggs. So, the benefit of mating repeatedly, in a general sense, is stronger for males than females.

Populations in which many males mate with more than one female are common. Of course, if the adult sex ratio is 1:1, this means that there are a lot of unmated males. There are at least four situations in which males have the opportunity to mate polygynously: if the male (i) can defend many females from other males, (ii) can defend access to a share of the resources required by females or their offspring, (iii) provides no parental care but competes indirectly with other males (scramble competition), or (iv) provides no parental care but competes directly with other males (lek behaviour).

To be clear, any type of polygynous mating system focuses on the male strategy for mating. Although polygyny is often defined as one male mating **exclusively** with more than one female, this exclusivity may be in appearance only. While females might have little to gain by copulating with other males, there are many instances when they do. The fact that extrapair copulations have been determined to occur relatively frequently suggests that both males and females may mate with more than one individual. So, polygyny technically happens when the male mates with more than one female, while the female may or may not mate with one or more males. With this in mind, let's look at the strategies for males while also noting what the females may be doing.

12.3.1 Female defence polygyny

Females often breed in groups. Most often, this is related to protection; groups of females, and perhaps associated males, offer better protection for themselves and their young. When females bunch up, it's easier for males to defend and breed with groups of them. Males often compete directly for this right, so size is typically beneficial.

Many seals breed on well-protected sites, and males provide little parental investment other than their sperm. Even though they may protect females from the advances of other males, this has little to do with the wellbeing of the female except for a possible respite from harassment. One of the best and most frequently used examples of a species in which males can defend groups of females (often referred to as harems) is the northern elephant seal (*Mirounga angustirostris*). Males arrive first at breeding sites (typically beaches) and compete amongst themselves, waiting for the arrival of females. Being large and strong is beneficial in these fights, and males are more than twice as large as females. Females arrive on beaches and give birth to a pup fertilized the previous year, and the winners of male-male contests will defend the most females. Females have little need to copulate with more than one male since they only produce a single pup. In one study, only eight of 123 females were observed copulating with two males, with the first being the genetic father in seven cases.[7] Thus, although males are driven to mate with as many females as possible, females stand little to gain by mating with more than one male.

In many cases, female fish that live on coral reefs group together, primarily because of habitat limitations, and can be economically defended by males. Lagoon triggerfish (*Rhinecanthus aculeatus*; Figure 12.3) demonstrate various mating systems, including polygyny, monogamy, and promiscuity. Females defend individual territories within the larger territories of males who discourage the advances of other males for up to five females.[8] Males don't provide parental care, so they can devote their time to courting females. Females don't appear disadvantaged by mating with a male who mates with other females. Still, monogamous relationships are more prevalent later in the breeding season as the number of available females declines.

Female defence polygyny is also observed in invertebrates. The ant genus *Cardiocondyla* consists of about 100 species native to Africa, Australia, and Eurasia that live in relatively (for ants) small groups. All species have wingless **ergatoid** males that may or may not live together with winged males. However, while winged males found in some species are capable of dispersing, ergatoid males remain in their natal colony and vigorously fight rival males for the right to breed. Generally, only one male remains and mates with any sexual females who emerge.[9] Thus, while the dominant male mates numerous times with a group of females, the females only mate once, receiving all the sperm they will need during their lives.

12.3.2 Resource defence polygyny

It's often easier to defend a patch of resources than a bunch of females, so resource defence polygyny is generally more common than female defence polygyny. Most people recognize an odonate as they fly around the edges of ponds. These are typically dragonflies but could be members of the closely related damselflies. The difference? Dragonflies are usually more robust and perch with their wings straight out from their bodies. Damselflies are often slenderer and perch with their wings touching above them. In many of these species, whether dragonflies or damselflies, males often defend a territory from other males and mate with any female that enters it. Are they protecting

Figure 12.3 *Fortunately, the lagoon triggerfish is relatively small because the male will vigorously defend his territory against all intruders, even human divers. Females defend smaller areas enabling male territories to encompass up to five females. Later in the season, as breeding females become increasingly rare, males may be limited to monogamous relationships.*
Credit: Minakryn Rusian/Shutterstock Photo ID 1287044611.

particular resources used by females? Sometimes, the males defend the sites just for display purposes; females visit and mate but don't need anything within the territory except for the male's sperm. However, male rubyspot damselflies (*Hetaerina rosea*) defend vegetation on which females lay their eggs. When the amount of vegetation is experimentally manipulated, the number of males and females decreases or increases accordingly.[10]

In some cases, there is a fine line separating female defence from resource defence polygyny, as it is often unclear if males defend the territory containing a resource needed by females or the females themselves. Neotropical harvestmen (*Acutisoma proximum*; the common type in North America is often called 'daddy long legs') males appear to switch from one to the other during the breeding season.[11] Early in the season, males defend territories containing oviposition sites (vegetation) to which females come, copulate and lay their eggs. So, males, at this point, are securing the resources needed for reproduction by females; the male that defends the most (or best) plants attracts the most females. Later, males switch from guarding the plants to defending females directly.

Green poison arrow frogs (*Dendrobates auratus*; Figure 12.4) demonstrate how mating systems are a trade-off between what is best for each sex. Males defend territories containing oviposition sites for females and benefit by mating with more than one

Figure 12.4 *Despite their name, green poison frogs come in a variety of colours. In all forms, however, males assume all parental care duties once the eggs have been laid. Typically, this involves protecting the eggs and tadpoles and carrying offspring to nearby tree hollows, where they undergo metamorphosis. Males can improve their reproductive performance by mating with more than one female. However, because of the male's extensive parental care, the success of each female declines as the male takes on more mates.*

Credit: Nino/Adobe Stock Photo ID 556835471.

female.[12] Females, on the other hand, maximize their fitness if the male mates only with them. Once the eggs are laid, males provide all the parental care—he looks after the eggs and then carries the tadpoles on his back when searching for tree holes filled with water where the young undergo metamorphosis. As the number of eggs increases, the male can't provide the same care per egg; thus, a female's cost is substantial if he mates repeatedly. As a result, they will act aggressively towards other females who come too close and **pseudocourt** the male to divert his attention from the intruder.

12.3.3 Scramble competition polygyny

Sometimes, defending either resources or females from other males is not feasible. That's because they may be widely dispersed, or conversely, the resources may be too concentrated and competition with other males would take too much energy. Scramble competition polygyny is probably the most common mating type among animals.

Generally, males approach females and, after a courtship of variable length, copulate with them. They may stick around and try to prevent other males from copulating with her, or they may move on quickly and attempt to breed with another female. Searching for receptive females is often an essential characteristic of scramble competition, often resulting in the evolution of traits that enable males to find females more quickly.

Thirteen-lined ground squirrels (*Spermophilus tridecernlineatus*) illustrate scramble competition polygyny's main features.[13] These squirrels are widely found in the open areas around central North America. They look similar to chipmunks, but their stripes (yes, 13 of them) are found only on the body and not the cheeks like chipmunks. Males typically search for sexually receptive females during the breeding season and show no signs of defending a territory. Although having other males around decreases the chance of male copulation, intra-male conflict is low and breeding, for the most part, is a peaceful affair. The females may breed with more than one male. In many rodents, significant sperm competition occurs, meaning that females mate repeatedly. The number of times is unclear, but males who copulate first are more likely to fertilize the eggs.

Scramble polygyny is widespread in insects and has been particularly well-studied in leaf beetles. Males and females of the beetle *Leptinotarsa undecimlineata* spend most of their lives on two particular host plants of the Solanaceae (nightshade or potato) family. When breeding, males travel from plant to plant, acting as sites of copulation and egg-laying for females. After mating with a female, a male often stays and guards her against the advances of other males but does not defend a territory or any group of females. Males are likelier to switch plants when the proportion of available females declines.[14] Females also change plants but tend towards those with fewer males, suggesting they may be avoiding male harassment. It's unclear if females benefit by mating more than once in this species.

12.3.4 Leks

I watched a comedy the other night, and two leading male characters entered a bar. They intended to perform certain behaviours that would make them more attractive to women and, if all went successfully, engage in sexual intercourse with them. Not on the dance floor, of course, but in private after a minimal amount of courtship. In general, these meeting places are where males go, sometimes alone and sometimes in groups, to increase their opportunities to meet and interact with members of the opposite sex. Presumably, some women do the same, although most tell me (and I have no reason not to believe them) they go there only to dance.

These bars are the human version of a lek breeding system, although there are certain differences. Both involve males gathering in a particular area and competing with each other, either directly or indirectly, to improve their chance of copulating. Leks are characterized by occurring in relatively small areas or arenas that contain no resources that females need, the absence of male parental investment (outside of sperm), the ability of females to assess and mate with certain males, and copulation in, or close to, the arena after the assessment is made. Typically, there is a substantial variation in male mating success, with only a small proportion of the males breeding with most of the females. Once they've received the sperm of a male, females leave the area and produce their offspring. I'll leave it to you to determine the properties that render the bar analogy somewhat inaccurate.

The sound of cicadas buzzing in trees always reminds me that the long, hot days of summer will soon be over. The periodic cicada (*Magicicada sp.*), of which there are seven species, lives underground, sucking on the sap from roots for (usually) 13 or 17 years before emerging to breed. During their three-week reproductive phase, males gather in trees and produce auditory signals to which females are attracted. (Many species of fireflies, especially in the tropics, have a similar system, but males use visual cues to entice females.) After a brief courtship lasting about a minute, individuals engage in sex for a few hours, and the females depart and lay their eggs. After copulation, females become unreceptive to other males and only mate once. Males, on the other hand, immediately begin their attempts to attract another female to mate. Many males mate polygynously, while most die without ever copulating.

Ruffs (*Calidris pugnax*) are lekking shorebirds that display a high degree of sexual dimorphism, with the male being much larger than the female. Up to about 20 males compete amongst themselves and hold tiny territories (about 1 m^2) on the lekking grounds. Females (called reeves) assess the males and mate before leaving the area, raising their brood alone. Interestingly, females will mate with several males and exhibit the highest proportion of mixed paternity broods for all avian lekking species.[15] So once again, both males and females mate with more than one partner during each breeding season and claiming it is 'lek polygyny' doesn't tell the whole story, at least from the female point of view.

12.4 Polyandry should be less common

Females experience both costs and benefits of mating with more than one individual during the breeding season, and these consequences differ for females of different populations or species. Some potential disadvantages of mating with more than one male are the increased time spent on mating, greater exposure to predators and parasites, and even physical damage in some species. Some of the potential advantages are insurance against partial or complete infertility, increasing the genetic variability of offspring, the encouragement of sperm competition, improved physical condition or increased longevity (if males provide nuptial gifts), and, in some cases, more care for offspring if more than one male contributes to their wellbeing. Females must (evolutionarily) weigh these costs and benefits to determine the optimal number of times to mate. In most cases, the number of partners will be lower than the optimum for males, so a strictly polyandrous mating system (i.e. females having exclusive relationships with more than one male) is expected to be rare. However, females often mate repeatedly, even when the breeding system is classified as polygynous.

The spotted sandpiper (*Actitis macularius*) is the most common sandpiper in North America and is often classified as polyandrous. Members of this species typically breed in areas where significant flushes of insects provide an abundant food supply. Females produce clutches of four eggs, and their own physiology precludes them from laying larger clutches. For this reason, if their condition permits them to lay more eggs, they must produce more clutches. The eggs need to be incubated for about three weeks, and although the chicks can feed themselves when they hatch, they need to be protected because they can't fly for another couple of weeks. If the female were to look after the offspring herself, she could only produce one nest a year. On the other hand, if her mate assumes all of the parental duties, she can use that extra energy to make another clutch of eggs. Therefore, she competes with other females for access to males. In some cases, a single female can produce up to four clutches of eggs for four different males.

In some birds, females may mate with more than one male if they can get them both to care for their offspring. Individual dunnocks (*Prunella modularis*), which we've looked at previously, can mate monogamously, polygynously, polyandrously, or polygynandrously (i.e. more than one male and more than one female). Although a female will only produce a single clutch of eggs, mating with two males encourages them both to care for her offspring, as neither can be sure how many of them are genetically his. How much they care is correlated with how much they copulate with the female, which correlates with the number of offspring they fathered.

Perhaps the most extreme polyandry occurs in the social hymenopterans. Recall that these insects are characterized by a haplodiploid genetic system, where the females are haploid, and the males are diploid. When the queen mates, she receives sperm from numerous males, who die after mating. In giant (*Apis dorsata*) and western honeybees (*Apis mellifera*), the queens have been observed to copulate with more than 100 and 45 males, respectively.[16] The reason for this high level of polyandry isn't entirely apparent, although various hypotheses have been put forward, including increasing the genetic heterogeneity of the colony.

Springbok mantids (*Miomantis caffra*; Figure 12.5) reproduce using both scramble competition polygyny and polyandry, illustrating why it is difficult to designate a population mating system. Females are cannibalistic, often consuming males, even without mating. To avoid this, males act aggressively towards females and may stab them in the abdomen with their claws during precopulatory foreplay. Despite the potential risk of injury, females will mate with up to three males.[17] Interestingly, some females avoid males altogether, producing offspring parthenogenetically, although their lifetime reproductive success tends to be lower than those who reproduce sexually. The mating strategies of females probably reflect trade-offs in costs and benefits. Females with fewer mates don't suffer as many costs; those with more mates maximize the benefits.

Figure 12.5 *Springbok mantids, seen here copulating, have rough sex. Females may cannibalize males, and males may stab females in the abdomen. Despite this, both sexes may copulate with more than one individual. Because of different mating possibilities in this and many other species, assigning a general mating pattern to a population is difficult.*

12.5 Alternate strategies

In many species, males or females may be unable to breed as often or effectively using the more typical route and are forced to use other methods to maximize their reproductive success. For example, if large males are better able to defend territories or groups of females, what can you do if you happen to be a reproductively mature but small male? Since you can't compete directly with other males, you might obtain copulations with females more surreptitiously. I don't want to go into much detail concerning the various strategies, but recent evidence suggests that both sexes, but mainly males, may use alternative approaches to maximizing their fitness. Let's look at alternate methods males use in two of the species mentioned previously.

Over 80% of the breeding male ruffs gather on leks as described above.[18] They compete directly in defined arenas, and the dominant males typically copulate with more females. However, there are two other types of males. Satellite males make up just less than 20% of the remaining males. They sit around the arena and are physically distinct and smaller than the territorial males. Their strategy is avoiding confrontations with territorial males and intercepting females who may be coming to copulate. Although satellite males copulate less often than expected for their numbers, their strategy may be low benefit and minimal risk, enabling them to live longer. Another rare type of male has been identified, one that appears to mimic females.[19] We are reasonably sure they are good mimics since the territorial males often attempt to copulate with them. Like satellite males, mimics try to steal copulations, but they can get closer to dominant males without arousing suspicion. To help them in their quest to fertilize eggs, they have testes 2.5 times the size of the dominant males'.

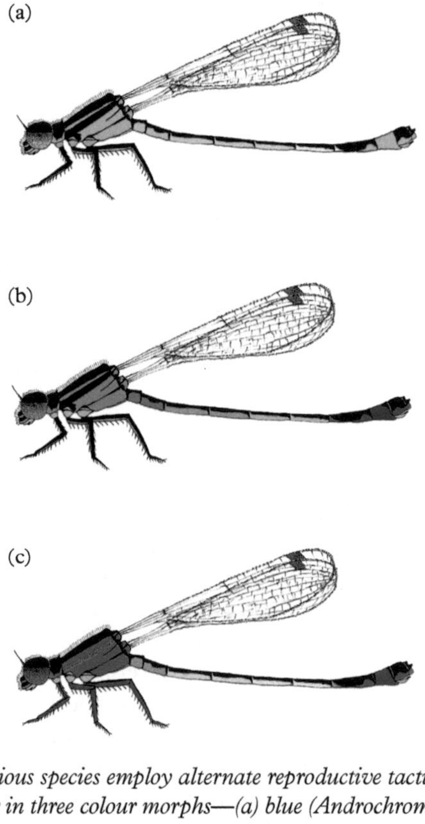

Figure 12.6 *Females of various species employ alternate reproductive tactics during breeding. Blue-tailed damselflies occur in three colour morphs—(a) blue (Androchrome), (b) green (Infuscans), and (c) orange (Infuscans-obsoleta). The three morphs differ behaviourally, including their resistance and tolerance to male mating attempts. The attractiveness to males of the three morphs is based partially on their frequency in the population.*

Alternate mating strategies may be used on a permanent or temporal basis. In many cases, males of different ages may use different mating strategies. This scenario is often true when males continue to grow, even after reaching sexual maturity. For example, younger northern elephant seal males, although sexually mature, are smaller and unable to compete with the more considerable 'harem masters'. A good strategy, therefore, is to mimic females. By looking like members of the opposite sex, they can easily infiltrate harems and attempt to copulate with females behind the dominant male's back.

Although alternate mating strategies seem less common in females, they certainly exist. In many species of damselflies, females exhibit two or more colour patterns. Although there have been many hypotheses concerning the function of this female polymorphism, recent studies have drawn a link between the colour morph and behaviour, suggesting that females exhibit alternate reproductive tactics. For example, all male blue-tailed damselflies (*Ischnura elegans*) are blue. However, females occur in three distinct colour morphs (Figure 12.6)—blue (androchrome morph), green (gynochrome morph), and orange (rare gynochrome morph). Males prefer green gynochrome females, even when similar numbers of blue morphs occur;[20] the offspring survival of blue morph females is, however, higher. This difference in offspring survivorship may compensate for their decreased attractiveness to males. An analogous situation occurs in common lizards (*Lacerta vivipara*), with females having ventral colouration ranging from yellow to bright orange. Differences in clutch size and hatching success are apparent for different colour morphs.[21] Further study on females of other species will undoubtedly unveil several reproductive strategies.

12.6 Summary

Mating systems are highly variable across the animal kingdom. They should be viewed not as population or species characteristics, but as common strategies of individual males and females designed to maximize their own reproductive fitness. In most cases, what is best for one sex will not be optimal for the other, so sexual conflict develops. Any patterns reflect trade-offs between these strategies, with both sexes trying to maximize the benefits and minimize the costs of mating sexually with another individual. Sometimes, only single individuals pair for one or more breeding bouts. This mating system is often called '**social monogamy**', but it's important to remember that both sexes may copulate with more than one member of the other sex. Polygynous and polyandrous mating systems may also contain a mixture of males and females that breed with one, more than one, or no individuals. Despite this, generalizations can be made as environmental opportunities increase individuals' chances of having more than one mate. For example, where resources required by females are clumped, males may defend these assets from other males, increasing the probability that they will mate with more than one female. Or when abundant food resources controlled by males enable females to produce more eggs, they may mate with more than one male. I discussed the various environmental conditions that may increase or decrease the chances of mating with more than one individual; still, in all cases, males and females are expected to make breeding decisions to maximize their lifetime fitness.

References

1. Kokita T, Nakazono A. Sexual conflict over mating system: The case of a pair-territorial filefish without parental care. Anim Behav [Internet]. 2001 Jul [cited 2022 Sep 3];62(10):147–155. Available from: https://doi.org/10.1006/anbe.2001.1738
2. Steinauer ML. The sex lives of parasites: Investigating the mating system and mechanisms of sexual selection of the human pathogen *Schistosoma mansoni*. Int J Parasitol [Internet]. 2009 Aug [cited 2022 Sep 3];39(10):1157–1163. Available from: https://doi.org/10.1016/j.ijpara.2009.02.019
3. Whiteman EA, Côté, IM. Social monogamy in the cleaning goby *Elacatinus evelynae*: Ecological constraints or net benefits? Anim Behav [Internet]. 2003 Aug [cited 2022 Sep 3];66(2):281–291. Available from: https://doi.org/10.1006/anbe.2003.2200
4. Lambert CT, Sabol AC, Solomon NG Genetic monogamy in socially monogamous mammals is primarily predicted by multiple life history factors: A meta-analysis. Front Ecol Evol [Internet]. 2018 Sep [cited 2022 Sep 3];6:139. Available from: https://doi.org/10.3389/fevo.2018.00139
5. Wojczulanis-Jakubas K, Jakubas D, Chastel O. Different tactics, one goal: Initial reproductive investments of males and females in a small Arctic seabird. Behav Ecol Sociobiol [Internet]. 2014 Jul [cited 2022 Aug 22];68:1521–1530. Available from: https://doi.org/10.1007/s00265-014-1761-4
6. Decker MD, Parker PG, Minchella DJ, Rabenold KN. Monogamy in black vultures: Genetic evidence from DNA fingerprinting, Behav Ecol [Internet]. 1993 Mar [cited 2022 Aug 23];4(1):29–35. Available from: https://doi.org/10.1093/beheco/4.1.29
7. Fabiani A, Galimberti F, Sanvito S, Hoelzel AR. Extreme polygyny among southern elephant seals on Sea Lion Island, Falkland Islands. Behav Ecol [Internet]. 2004 Nov [cited 2022 Aug 23];15(6):961–969. Available from: https://doi.org/10.1093/beheco/arh112
8. Ziadi-Künzli F, Tachihara K. Female defence polygyny and plasticity in the mating system of the demersal triggerfish *Rhinecanthus aculeatus* (Pisces: Balistidae) from Okinawa Island. Mar Biol [Internet]. 2016 Jan [cited 2023 Jun 15];163:27. Available from: https://doi.org/10.1007/s00227-015-2780-z
9. Heinze J, Hölldobler B. Insect harem polygyny—The case of *Cardiocondyla* ants: A comment on Griffin et al. Behav Ecol Sociobiol [Internet]. 2019 Jun [cited 2022 Aug 24];73:99. Available from: https://doi.org/10.1007/s00265-019-2718-4
10. Guillermo-Ferreira R, Del-Claro K. Resource defense polygyny by *Hetaerina rosea* Selys (Odonata: Calopterygidae): Influence of age and wing pigmentation. Neotrop Entomol [Internet]. 2011 Feb [cited 2022 Aug 24];40(1):78–84. Available from: https://doi.org/10.1590/s1519-566x2011000100011
11. Buzatto BA, Machado G. Resource defense polygyny shifts to female defense polygyny over the course of the reproductive season of a Neotropical harvestman. Behav Ecol Sociobiol [Internet]. 2008 Aug [cited 2022 Aug 25];63(1):85–94. Available from http://www.jstor.org/stable/40645496
12. Summers K. Sexual conflict and deception in poison frogs. Curr Zool [Internet]. 2014 Feb (cited 2022 Aug 25];60(1):37–42. Available from: https://doi-org.proxy.library.upei.ca/10.1093/czoolo/60.1.37

13. Schwagmeyer PL, Woontner SJ. Scramble competition polygyny in thirteen-lined ground squirrels: The relative contributions of overt conflict and competitive mate searching. Behav Ecol Sociobiol [Internet]. 1986 Nov [cited 2022 Aug 25];19(5):359–364. Available from: https://doi.org/10.1007/BF00295709
14. Muniz DG, Baena ML, Macías-Ordóñez R, Machado G. Males, but not females, perform strategic mate searching movements between host plants in a leaf beetle with scramble competition polygyny. Ecol Evol [Internet]. 2018 May [cited 2022 Aug 26];8(11):5828–5836. Available from: https://doi.org/10.1002/ece3.4121
15. Lank DB, Smith CM, Hanotte O, Ohtonen A, Bailey S, Burke T. High frequency of polyandry in a lek mating system. Behav Ecol [Internet]. 2002 Mar [cited 2023 Jun 15];13(2):209–215. Available from: https://doi.org/10.1093/beheco/13.2.209
16. Kraus FB, Neumann P, van Praagh J, Moritz RF. Sperm limitation and the evolution of extreme polyandry in honeybees (*Apis mellifera* L.). Behav Ecol Sociobiol [Internet]. 2004 Nov [cited 2023 Jun 15];55:494–501. Available from: https://doi.org/10.1007/s00265-003-0706-0
17. Burke NW, Holwell G. Costs and benefits of polyandry in a sexually cannibalistic mantis. J Evol Biol [Internet]. 2022 Dec [cited 2023 Jun 15];36(2):412–423. Available from: https://doi.org/10.1111/jeb.14135
18. Widemo F. Alternative reproductive strategies in the ruff, *Philomachus pugnax*: A mixed ESS? Anim Behav [Internet]. 1998 Aug [cited 2023 Jun 15];56(2):329–336. Available from: https://doi.org/10.1006/anbe.1998.0792
19. Jukema J, Piersma T. Permanent female mimics in a lekking shorebird. Biol Lett [Internet]. 2006 Jan [cited 2023 Jun 15);2(2):161–164. Available from: https://doi.org/10.1098/rsbl.2005.0416
20. Subrero E, Pellegrino I, Cucco M. Different stress from parasites and mate choice in two female morphs of the blue-tailed damselfly. Evol Ecol [Internet]. 2021 Aug [cited 2023 Jun 16];35:687–704. Available from: https://doi.org/10.1007/s10682-021-10130-z
21. Vercken E, Massot M, Sinervo B, Clobert J. Colour variation and alternative reproductive strategies in females of the common lizard *Lacerta vivipara.* J Evol Biol [Internet]. 2007 Jan [cited 2023 Jun 16];20(1):221–232. Available from: https://doi.org/10.1111/j.1420-9101.2006.01208.x

Summary

The evolutionary transition from prokaryotes to eukaryotes, which occurred about 2.5 billion years ago, was accompanied by a change in reproductive mode. Why was it that sexual reproduction became important at that time? Indeed, asexual reproduction was used in passing on genes for over a billion years and presently persists in all the prokaryotes and some of the eukaryotes. How sexual reproduction became so important and why the sexual strategies of males and females became so different are the primary themes in this book.

Nearly all organisms on Earth reproduce sexually, but many can pass on their genes asexually. Such methods include fission, budding, fragmentation, sporulation, and other forms often observed today. When compared with asexual reproduction, sexual reproduction is costly. It is energetically demanding and takes more time. In addition, it exposes individuals to increased parasitism and predation. Even without these costs, usually half the progeny produced by sexual reproducers can't generate any potentially reproductive offspring. This major cost is often referred to as the two-fold cost of males and remains the most significant drawback of sexual reproduction. There must have been one or more benefits to sexual reproduction if it were to evolve as the primary mode of reproducing in multicellular organisms.

But even if we assume that sexual reproduction provided some benefit, it still isn't clear why two very different sexes evolved. Males and females differ in various ways, depending on the species, but share one characteristic across all species: females produce a few large sex cells or gametes, while males produce many more small gametes. Furthermore, the gametes produced by females are usually stationary, while those produced by males can move and typically search out the female sex cells. Sure, a few species are isogamous, making two (sometimes more) types of gametes of similar size. But generally, nearly all species that reproduce sexually have males and females who combine gametes during reproduction. With the aid of models, it has been shown that two sex cells are more likely to come together if one is large and stationary and the other is small and mobile. Thus, when discussing the strategies males and females use, we must understand the constraints imposed upon each by gamete size and mobility.

So, what advantage did sexual reproducers have over those who reproduced asexually? One possibility is that meiosis, found in all sexually reproducing organisms today, evolved as a response to organisms' increased use of oxygen in producing energy. While using oxygen made for a more efficient metabolism, it also had a dark side: the free radicals produced during oxidative phosphorylation increased DNA damage. Combining your genetic material with a closely related individual was a way of getting undamaged stretches of DNA that could be combined with yours to eliminate the damage or, at least, mask it. While such a process is observed in prokaryotic transformation, without reproduction taking place, sexual reproduction was simply an extension of this process in eukaryotes, increasing the likelihood that the germ line was relatively free of serious mistakes.

In addition (or alternatively), combining genes with those of another individual increased the variability of offspring. This variation became more pronounced with recombination and the random segregation of alleles during meiosis. By creating differences among offspring, selection would favour the most appropriate individuals(s) for that environment. As asexual reproduction can result in significantly more potentially reproductive offspring being produced in each generation, as well as avoiding many of the costs associated with sex, sexual reproduction had to provide substantial immediate benefits in each generation to evolve. A great deal of investigation has focussed on the value of producing offspring that can better combat the parasites that may have become well-adapted to their mothers. This "Red Queen Hypothesis" is currently regarded as one of the best explanations for the main advantage of sexual reproduction.

Males and females were already subject to different selection pressures when breeding, brought about by differences in gamete size. Over time, differences between the two sexes have often become magnified to the extent that we can't look at a typical breeding strategy of the species but rather individual breeding strategies of the two sexes. Charles Darwin stated that males compete for choosy females. Typically, this statement is true, as males can maximize their fitness by repeatedly mating. Sperm is cheap, and males usually produce it in abundance; as a result, they don't have to be that selective in their fertilization partner. Conversely, eggs are expensive, and females are more often limited by the number of gametes or offspring they can produce. For this reason, they need to be more selective about with whom they combine their genes.

Of course, sometimes females compete for choosy males, or both sexes are choosy. How long an individual is out of the breeding pool is most important as this determines their availability to mate. Generally, females are tied up in breeding duties longer than males; thus, males compete for the females who are accessible. In many species, however, the situation is reversed, and females may compete for the available males. Such conditions often exist when males, rather than females, spend extended periods caring for offspring. In other species, both sexes can benefit by being choosy about their mates, mainly when males and females both contribute substantially to parental care.

Males and females clearly have their own best interests in mind when they reproduce and always try to maximize their fitness. Maximizing fitness often involves maximizing the number of offspring sharing their genes in the next generation. But not always. In the long run, producing fewer high-quality offspring might be better than many low-quality ones. Regardless, males and females often use different strategies to reproduce. Conflict between the sexes often arises; what's best for the fitness of one sex may not be good for the fitness of the other. In some cases, strategies by one sex to maximize fitness can actually be detrimental to the other. Cases of sexual conflict between males and females of various species are well documented in the literature.

However, I don't wish to leave you with the impression that males and females wage a continual battle when producing shared offspring, as this would be misleading. By working together, both sexes can often produce more high-quality progeny than they could by themselves. In other words, the use of the other sex in the sharing of parental duties can increase one's fitness. Thus, we have cooperation as well as conflict between the sexes. It's important to note, however, that males and females cooperate because it serves their best interests, not because they have any desire to please their partner.

The individual examination of male and female strategies during sex renders it challenging to assign a particular mating system to a population. Individuals always face a unique situation when their reproductive season begins. What breeding individuals are available? Would they provide good-quality genes to mix with your own? How reliable would they be if they were needed to provide parental care? There are a multitude of such questions, and they make every breeding opportunity distinctive. Males and females often have multiple mating partners, although this may be for very different reasons. Males usually mate with more than one female because they are most often limited by the availability of receptive females (and their eggs). Females can mate more than once for several reasons, including hedging their bets against insufficient sperm, getting resources that are provided with copulations, improving the quality of their offspring, assessing their mate, and ensuring their breeding partners of paternity.

For these reasons, true genetic monogamy, in which one male and female mate, is relatively rare. Polygyny is more common, especially if our definition allows females to mate repeatedly, although perhaps surreptitiously. And polyandry, as defined as the exclusive mating between one female and more than one male, is also rare, especially if the breeding of males is limited to a single female. Thus, each population of organisms may have a variety of breeding practices, all depending on the biotic and abiotic environment individuals face when they mate.

Understanding the history of life on Earth has relied substantially on fossils. Unfortunately, most aspects of sexual reproduction don't fossilize, and we've had to rely on other means to reveal its possible origins. Using techniques developed over the past few decades, we can directly examine molecular characteristics of modern-day organisms, including gene sequences and their resulting proteins. These methods have been instrumental in determining phylogenetic histories. However, we often use educated hypotheses to suggest the evolution of certain traits. Like all scientific hypotheses, these are tested, usually using field and lab studies of living organisms, and are continually modified based on their support or lack thereof. Although determining the current costs and benefits of sexual reproduction is crucial to this understanding, it must be emphasized that recent studies may not provide suitable information to state definitively why it originally evolved. Such insight remains hypothetical, but a better understanding of why it is the primary form of reproduction today goes a long way to solving this riddle.

On September 22, 2022, tropical storm Fiona battered the small, rural island where I live. Fortunately, I wasn't in the country at the time, and the only damage to my personal property involved the loss of half of the soffit ceiling above the outside porch. Given the extensive property and coastal damage from the high winds and heavy rainfall in other regions of the island, we got off easy. As usual, I procrastinated fixing the ceiling, and a pair of robins decided the area was an excellent nesting site. The damaged ceiling opened up a ledge perfect for a robin nest, and the female (at least I assume it was the female) began building. Knowing my wife would not enjoy a robin nesting above the patio furniture, with the accompanying mess produced by the offspring, I worked quickly to fix the ceiling and eliminate the breeding site. This procedure, regrettably, involved continually removing any partially built nest. The female robin was persistent but had to give up when I fully enclosed the ceiling a few days later. The female left the immediate area, but the male stayed and attempted to attract another mate. He sang from 4:30 am to 9:30 pm for weeks, presumably to a point when all females had bred, and any offspring had fledged from nests. Every day, my wife and I listened to him, wishing he'd pair up both because he was making such an effort and also so we could get some peace. He never did find a mate, but presumably, if still alive, he will try again next year.

Sexual reproduction and the drive to pass on one's genes is all-encompassing, and the lives of all organisms revolve around it. In some species, individuals can only breed for a few hours, while others may produce repeated batches of offspring over many years. Regardless of individuals' time on Earth, the ability to reproduce is an integral part of their life cycle. Those who reproduce sexually need to overcome many obstacles that are not a detriment to asexual organisms, and males and females are often faced with different challenges. In many cases, this results in conflict between the sexes. However, their goal is the same – to pass on their traits to individuals of the next generation.

Index